4th International Conference on Mechanical Instrumentation and Automation

Journal of Physics: Conference Series Volume 2561

Chengdu, China
7-9 April 2023

ISBN: 978-1-7138-7759-2
ISSN: 1742-6588

Printed from e-media with permission by:

Curran Associates, Inc.
57 Morehouse Lane
Red Hook, NY 12571

Some format issues inherent in the e-media version may also appear in this print version.

Institute of Physics
Dirac House, Temple Back
Bristol BS1 6BE UK

Phone: 44 1 17 929 7481
Fax: 44 1 17 920 0979

techtracking@iop.org

Additional copies of this publication are available from:

Curran Associates, Inc.
57 Morehouse Lane
Red Hook, NY 12571 USA
Phone: 845-758-0400
Fax: 845-758-2633
Email: curran@proceedings.com
Web: www.proceedings.com

TABLE OF CONTENTS

Journal of Physics: Conference Series **2561** (2023) 011001 doi:10.1088/1742-6596/2561/1/011001

Preface

We are pleased to present the Proceedings from the 2023 4th International Conference on Mechanical Instrumentation and Automation (ICMIA 2023). The Conference took place in Chengdu, China during April 7th to 9th, 2023 (virtual event). It is the fourth Workshop of ICMIA series.

The Proceedings of ICMIA 2023 cover a broad range of topics concerned with research in mechanical instrumentation and automation: Instruments Science and Technology, Mechanical and Instrument Design, Intelligent Structural System, Detection and Control Technology, Automatic Detection and Instrumentation, Fault Diagnosis and Maintenance Theory, etc.

The Conference was organized as usual with keynote speeches, oral presentations, poster presentations, and academic investigation, attracting about 60 delegates around the world. The technical program started on the afternoon of April 7, with keynote speeches by various renowned professors on the next day morning. Dr. Samson S. YU (Deakin University, Australia) gave an address on Smart-Grids – An Overview. Next, Assoc. Prof. Ata Jahangir Moshayedi (Jiangxi University of Science and Technology, China) talked on From Electronic Nose to Gas Sensor Modelling. And lastly, Assoc. Prof. Kaixuan Chen (Aalborg University, Denmark) made a report on Deep Learning in Sensor-based Human Activity Recognition. And followed by the keynote speeches were several oral presentations given by students or researchers from universities or institutions worldwide. The Conference was closed with an academic investigation on the afternoon of April 9.

The Conference was indeed very successful! Internationally leading scientists have gathered in the ICMIA 2023 in various workshops and have created a stimulating scientific atmosphere for themselves and for the young scientists. ICMIA 2023 promoted the development and implications in the related fields of mechanical instrumentation and automation, and promoted the industrial cooperation of academic achievements.

We would like to thank everybody very much who contributed to the success of ICMIA 2023. We would like specifically to thank all the speakers, organizers, and the conference secretary, who helped in various ways and contributed in a very significant manner to the success of ICMIA 2023. In addition, we would like to thank the members of Journal of Physics: Conference Series, whose contribution made possible the volume.

The Committee of ICMIA 2023

Journal of Physics: Conference Series **2561** (2023) 011001 doi:10.1088/1742-6596/2561/1/011001

Committee Member

Conference Chair
Prof. Dongsheng Yu, China University of Mining and Technology, China
Prof. Hartmut Hinz, Frankfurt University of Applied Sciences (formerly FH Frankfurt), Germany
Prof. YUSRI YUSOF, Universiti Tun Hussein Onn Malaysia (UTHM), Malaysia

Technical Program Committee Chair
Prof. S.-H. Gary Chan, Hong Kong University of Science and Technology, China

Publication Chair
Assoc. Prof. Muhammad Junaid, China University of Mining and Technology, China

Technical Program Committee
Prof. Intan Zaurah Mat Darus, Universiti Teknologi Malaysia, Malaysia
Prof. Ian McAndrew, Capitol Technology University
Prof. Miadreza Shafie-khah. University of Vaasa, Finland
Prof. Xianbin Wang, Canada Research Chair at Western University, Canada
Prof. Ramash Kumar K. University of Madras, India
Prof. Angelo Cangelosi, University of Manchester, UK
Prof. Himani Goyal Sharma, Chandigarh University, India
Assoc. Prof. Hongze Wang, Shanghai Jiao Tong University, China
Assoc. Prof. Dr. M. Balakrishnan, M. Kumarasamy College of Engineering, India
Assoc. Prof. Taza, City University of Science and Information Technology, Pakistan
Assoc. Prof. Valerian Croitorescu, University Politehnica of Bucharest, Romania
Assoc. Prof. Saida (Boukhris) Gtifa, University of Carthage, Tunisia
Dr. Norfariza Ab Wahab, Universiti Teknikal Malaysia Melaka (Utem), Malaysia
Dr. Jyoti Prakash Panda, Dit University, India
Prof. Ozgur Koray, Biruni University, Turkey
Prof. Jelena Misic, Ryerson University, Canada
Prof. Surjeet Dalal, Teerthanker Mahaveer University, India
Assoc. Prof. Noor Zaman Jhanjhi, Taylors University, Malaysia
Assoc. Prof. Antti Rasila, Guangdong Technion – Israel Institute of Technology, Finnish
Assoc. Prof. Pinial Khan, Sindh Agriculture University Tando Jam Pakistan, Pakistan
Assoc. Prof. Asad Khattak, Zayed University, United Arab Emirates
Assoc Prof. Chawki Djeddi, Larbi Tebessi University, Algeria
Assoc Prof. Ruchi Doshi, University Azteca, Mexico
Assoc. Prof. Pardeep Arora, Head Department of Artificial Intelligence and Data Science at Kanya Maha Vidyalaya, India

Organizing Committee
Prof. Minoru Sasaki, Gifu University, Japan
Prof. Katarina Monkova, Technical University of Kosice, Slovakia
Prof. Intan Zaurah Mat Darus, Universiti Teknologi Malaysia, Malaysia
Prof. Mazhar Shaikh, Nizwa College of Technology, Oman
Prof. Heba Shaban, Technology and Maritime Transport Arab Academy for Science, Egypt
Assoc. Prof. Esmail Lakzian, Hakim Sabzevari University, Iran
Assoc. Prof. Ali H. Wheeb, University of Baghdad, Iraq
Assis. Prof. Neeraj Verma, Shiva Institute of Engineering & Technology, India
Assis. Prof. SYED ABDUL WAJID, National University of Science & Technology, Karachi
Dr. Norfariza Ab Wahab, Universiti Teknikal Malaysia Melaka, Malaysia

Peer Review Statement

All papers published in this volume have been reviewed through processes administered by the Editors. Reviews were conducted by expert referees to the professional and scientific standards expected of a proceedings journal published by IOP Publishing.

- **Type of peer review:** Single Anonymous
- **Conference submission management system:** Morressier
- **Number of submissions received:** 75
- **Number of submissions sent for review:** 71
- **Number of submissions accepted:** 36
- **Acceptance Rate (Submissions Accepted / Submissions Received × 100):** 48
- **Average number of reviews per paper:** 2
- **Total number of reviewers involved:** 15
- **Contact person for queries:**
  Name: Xuexia Ye
  Email: xx.ye@keoaeic.org
  Affiliation: AEIC Academic Exchange Information Centre

Journal of Physics: Conference Series **2561** (2023) 012001 doi:10.1088/1742-6596/2561/1/012001

Calibration of Heat Distortion and Vicat Softening Temperature Apparatus

Wei Gong[1], Wen Zhang[1], Lingjian Zeng[1], Tingting Ren[1*], and Cheng Zeng[1]

[1] Chongqing Academy of Metrology and Quality Inspection, Chongqing, 401123, China

[*]Corresponding author's e-mail: 755245093@qq.com

Abstract. Heat distortion and Vicat softening temperature apparatus are mainly used for measuring the thermal deformation temperature and Vicat softening point temperature of non-metallic materials. Parameters such as temperature indication error, heating rate error, deformation indication error, and mass indication error of loaded weights can directly decide the quality of heat distortion and Vicat softening temperature apparatus. There is no relevant calibration specification or method for heat distortion and Vicat softening temperature apparatus in China. This article presents a novel calibration method by determining corresponding environmental conditions, calibration standards, and calculation equations. The calibration results show that the method is scientific and practical for metrological performance evaluation of heat distortion and Vicat softening temperature apparatus.

1. Introduction

Heat distortion and Vicat softening temperature apparatus is mainly used for measuring the thermal deformation temperature and Vicat softening point temperature of non-metallic materials such as plastics, hard rubber, nylon, electrical insulation materials, long fibre reinforced composites, high-strength thermosetting laminated materials, etc. It is a highly intelligent testing instrument with accurate control performance and is widely used in colleges and universities, scientific research institutions and product quality supervision and inspection institutions [1-3].

The apparatus mainly comprises an indenter, heating equipment, weights, and a temperature-measuring device, as demonstrated in Figure 1. The indenter is made of hardened steel, at least 2 mm long, and fixed at the bottom of the rod. The heating equipment shall be a heating bath containing a suitable liquid. The heating equipment shall be provided with a control unit so that the temperature is raised at a uniform rate. A set of weights is used as the load applied to the test specimen. The heating bath is equipped with a separate temperature-measuring device at each test station. In this case, the temperature-sensing part of the instrument shall be located not farther than 12.5 mm from the point where the indenting tip contacts the specimen. The temperature-sensing part of the instrument shall not touch the specimen or be in contact with any part of the frame [4-6].

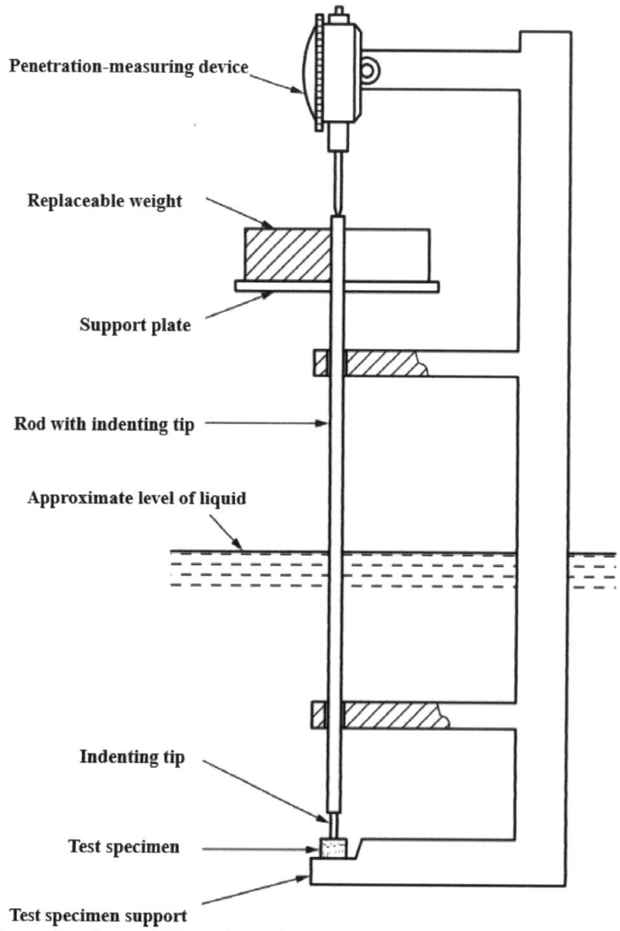

Figure 1. Schematic view of Heat distortion and Vicat softening temperature apparatus.

Heat distortion and Vicat softening temperature apparatus test the heat distortion characteristics of non-metallic materials. The accuracy of indicated parameters such as temperature, heating rate, deformation, and mass of weight can greatly influence the quality assessment conclusion of the tested material. However, there is no calibration specification for heat distortion and Vicat softening temperature apparatus in China at present, and the metrological performance of the apparatus cannot be ensured. Therefore, it is urgent to develop an effective calibration method that can evaluate the metrological performance of heat distortion and Vicat softening temperature apparatus to ensure the reliability of the apparatus in daily use.

2. Calibration conditions

2.1. Environmental conditions
During the whole measurement process, the temperature should be (18~28) °C, and the humidity should be no more than 80%RH.

2.2. Metrological standards
Temperature itinerant detecting instrument whose maximum permissive error is ±0.1 °C. A platinum resistance thermometer of A grade or better, and an electronic stopwatch whose maximum permissive error is ±0.5 s/d, are needed to calibrate the heating rate of the apparatus. A metallic gauge block of 5

class or better is needed to calibrate deformation. Electronic balance of Ⅲgrade or better is needed to calibrate the mass of the loaded weights.

3. Calibration method

3.1. Temperature indication error

The temperature itinerant detecting instrument is preheated for more than 30 min before starting calibration. The platinum resistance thermometer is inserted in the test hole of the sample rack of the heat distortion and Vicat softening temperature apparatus (if there are multiple sample racks, a platinum resistance thermometer is placed in each sample rack) and the thermometer is fixed [7-8]. The position of the platinum resistance thermometer and the tested sample shall be in the same position. The connection diagram of temperature indication error calibration is demonstrated in Figure 2.

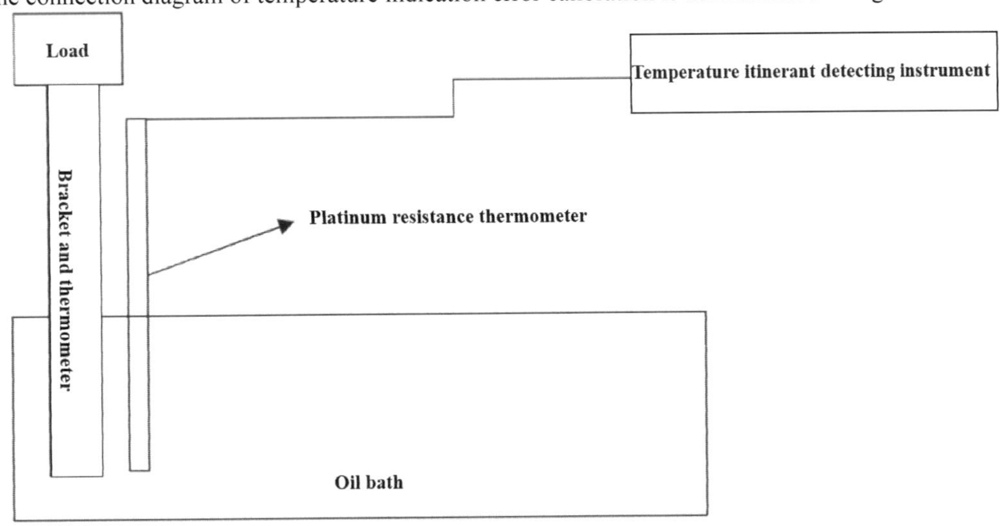

Figure 2. Connection diagram of temperature indication error calibration

The heat distortion and Vicat softening temperature apparatus are turned on. The upper limit temperature and heating rate of the apparatus is set. The deformation zero position is adjusted until the requirements are met. The mixing device of the apparatus is started. The temperature indication value of the temperature itinerant detecting instrument t_s and the temperature indication value of the apparatus t_i are recorded every 10 °C from 40 °C until the temperature of the oil bath reaches the set upper limit temperature. The temperature indication error of the apparatus Δt is calculated according to Equation (1).

$$\Delta t = t_i - t_s \tag{1}$$

The maximum of all calculated results is taken as the temperature indication error of the apparatus.

3.2. Heating rate error

The calibration of the heating rate error and temperature indication error shall be carried out simultaneously. As soon as the heat distortion and Vicat softening temperature apparatus are started, we read the initial temperature value of the temperature itinerant detecting instrument v_0 and start timing; we read the indication value of the temperature itinerant detecting instrument v_e again after 1 hour [9-10]. The measured heating rate v_s is calculated according to Equation (2).

$$v_s = \frac{v_e - v_0}{1} \tag{2}$$

The heating rate error Δv is calculated according to Equation (3).

$$\Delta v = v_d - v_s \tag{3}$$

Journal of Physics: Conference Series **2561** (2023) 012001 doi:10.1088/1742-6596/2561/1/012001

where v_d is the nominal value of the heating rate of the heat distortion and Vicat softening temperature apparatus.

3.3. Deformation indication error

The deformation indication value of each sample rack of the heat distortion and Vicat softening temperature apparatus is adjusted to zero. The deformation indication error of the apparatus is calibrated with a gauge block. The deformation indication error Δl is calculated using Equation (4).

$$\Delta l = l_i - l_s \tag{4}$$

where l_s is the nominal value of the gauge block, and l_i is the deformation indication value of the apparatus.

The maximum of all calculated results is taken as the deformation indication error of the apparatus.

3.4. Mass indication error of loaded weights

The electronic balance is preheated for more than 30 min. The self-calibration of the balance is completed before calibration. The electronic balance is used to weigh the load weights of the heat distortion and Vicat softening temperature apparatus one by one, and the mass of each load weight m_i shall be recorded. The mass indication error of the apparatus Δm is calculated according to Equation (5).

$$\Delta m = m_i - m_s \tag{5}$$

where m_s is the nominal value of the weight.

The maximum of all calculated results is taken as the mass indication error of the loaded weights.

4. Experimental verification of calibration method

5 typical types of heat distortion and Vicat softening temperature apparatus are selected as the calibration subjects for the verification experiment. The calibration results are demonstrated in Table 1.

Table 1. Calibration results of 5 typical types of heat distortion and Vicat softening temperature apparatus.

Type	Parameter	Result
XWB-300E	Temperature indication error	0.2 °C
	Heating rate error	0.3 °C/h
	Deformation indication error	10 μm
	Mass indication error of loaded weights	0.50 g
WKST-A	Temperature indication error	0.1 °C
	Heating rate error	-0.5 °C/h
	Deformation indication error	12 μm
	Mass indication error of loaded weights	-0.40 g
HDT Vicat 4U	Temperature indication error	0.2 °C
	Heating rate error	-0.2 °C/h
	Deformation indication error	-11 μm
	Mass indication error of loaded weights	0.60 g
BWR-300A	Temperature indication error	-0.4 °C
	Heating rate error	0.5 °C/h
	Deformation indication error	15 μm
	Mass indication error of loaded weights	-0.50 g
HS-XRW-300HA	Temperature indication error	-0.4 °C
	Heating rate error	0.2 °C/h
	Deformation indication error	-17 μm
	Mass indication error of loaded weights	0.70 g

As we can see from Table 1, the maximum temperature indication error is no more than ±0.4 °C, the maximum heating rate error is no more than ±0.4 °C/h, the maximum deformation indication error is no more than ±17 μm, which can meet the requirement of the enterprise standards of manufacturers.

The maximum indication error of loaded weights is no more than ±0.70 g, which can meet the requirements of JJG 99-2006 *Weights* and the enterprise standards of manufacturers. The feasibility and applicability of the calibration method have been proven.

5. Discussion and analysis

Methods of calibration of the temperature indication error include static and dynamic calibration, as elaborated in ISO 306 *Plastics - Thermoplastic Materials – Determinations of Vicat Softening Temperature (VST)*. The calibration method presented in the former chapters is designed for dynamic calibration, which can meet the needs of daily calibration by comparing the temperature rising curve measured by the temperature itinerant detecting instrument and platinum resistance thermometer. The temperature rising curve of the temperature indication value of heat distortion and Vicat softening temperature apparatus is demonstrated in Figure 3.

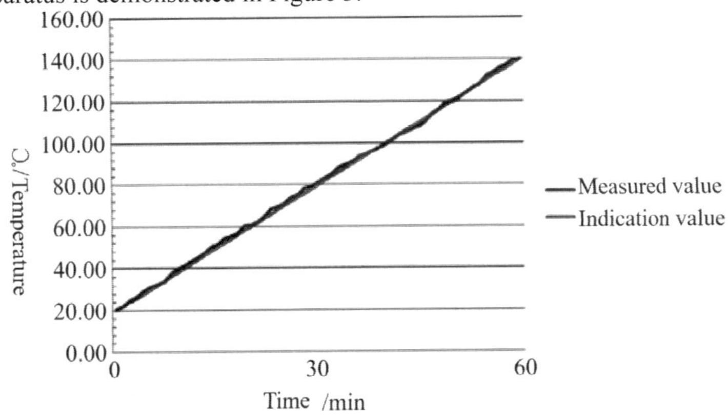

Figure 3. Comparison of temperature rising curves

If the temperature indication error at a certain temperature point is unqualified, a static calibration method shall be needed to precisely determine the correction value to correct the heat distortion and Vicat softening temperature apparatus. The static calibration method can be illustrated in Figure 4.

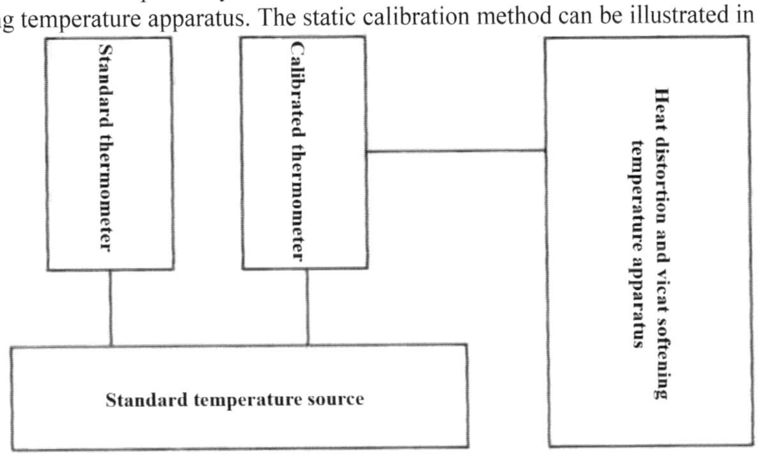

Figure 4. Schematic view of static calibration

However, static calibration has low efficiency and is inconvenient to carry to the site for daily calibration. Therefore, static calibration can be used as a complementary method for dynamic calibration, but the dynamic calibration method is still the main calibration means of temperature indication error.

6. Conclusion

This article presents a novel method for heat distortion and Vicat softening temperature apparatus to calibrate temperature indication error, heating rate error, deformation indication error, and mass indication error of loaded weights of the apparatus by determining corresponding environmental conditions, calibration standards, and calculation equations. The calibration results show that the method is scientific and practical for metrological performance evaluation of heat distortion and Vicat softening temperature apparatus. The follow-up study will develop a specific calibration device for heat distortion and Vicat softening temperature apparatus. The device will integrate the functions of the thermometer, gauge block, and balance to reduce the size of the measurement apparatus, improve its portability, and upgrade the calibration scheme and accuracy.

References

[1] Gong, W, et al. (2018) Research on calibration method of temperature for the thermal deformation vicat softening point temperature tester [J]. Metrology & Measurement Technique, 45(09): 83-85.
[2] Fu, K, et al. (2016) On-line calibration method for temperature parameters of thermal deformation and vicat softening point temperature tester [J]. Shanghai Measurement and Testing 2: 29-31.
[3] Wang, H. (2023) Study on the calibration method of thermal deformation vicat softening temperature tester [J]. Brand & Standardization 1: 56-58.
[4] Yao, J, R, et al. (2022) Verification and analysis of measuring ability of vicat softening temperature of plastics [J]. Machinery Industry Standardization & Quality 8: 27-29.
[5] Li, Y, X, et al. (2020) Research on vicat softening point temperature measurement method and Instrument [J]. Mechanical & Electrical Engineering Technology 49(08):91-92.
[6] Wu, D L, et al. (2018) Study on proficiency testing for measurement of vicat softening temperature [J]. Plastic Science and Technology 46(10):108-111.
[7] ISO. (2020) ISO 75-1 Plastics — Determination of temperature of deflection under load — Part 1: General test method [S]. ISO copyright office, Geneva.
[8] Xie, P, et al. (2019) GB 1634.1 Plastics — Determination of temperature of deflection under load — Part 1: General test method [S]. Standards Press of China, Beijing.
[9] ISO. (2022) ISO 306 Plastics — Thermoplastic materials — Determination of Vicat softening temperature (VST) [S]. ISO copyright office, Geneva.
[10] Xu, Y, N, et al. (2000) GB 1633 Plastics — Thermoplastic materials — Determination of Vicat softening temperature (VST) [S]. Standards Press of China, Beijing.

Comprehensive Improvement and Application of Copper Row Chain System of TOBSPIN Tobacco Cutter

Min Xu, Li Xu, Jianqiang Gong, Kelei Sun *, Yue Long, Qiujing Han, Miao Wang

Ningbo Cigarette Factory, China Tobacco Zhejiang Industry Co., Ltd., Ningbo 315040, China

* Corresponding author: 417612486@qq.com

Abstract: In order to solve the problem of abnormal wear of the lower copper row chain of TOBSPIN tobacco cutter, the working principle of the compaction unit of the wire-cutting machine was analyzed. Comprehensive improvements were made to the lower copper row chain system, which effectively reduced the wear on both sides of the copper row chain and prolonged the usage time of the copper row chain. Tested with TOBSPIN shredder used by Ningbo Cigarette Factory, the results showed that the wear speed of the improved copper exhaust chain was reduced from 2.62 mm / 800 h to 0.85 mm / 800 h, down by 67.6%. This method can support improving the service life of TOBSPIN tobacco cutters.

1. Introduction

The shredder is a special equipment for cutting the pre-treated tobacco leaves or stems into shredded tobacco or stems. It has the highest technical content and the most complex mechanical structure in silk-making equipment[1]. TOBSPIN tobacco cutter is a new type of rotary shredder produced and developed by HAUNI Company of Germany, which has the advantages of high automation, simple operation, and low noise[2]. It can meet the requirements of flexible and modular processing of cigarettes. In the daily use of maintenance, the disassembly of the copper strip chain is difficult for the TOBSPIN tobacco cutter, taking a long time and energy to separate the chain buckle hinge. The gap between both sides of the lower copper strip chain and the frame gradually increases, and there is abnormal wear, which reduces the service life of the copper strip chain.

At present, there are many studies on TOBSPIN tobacco cutters. Among them, Zhang Zhigang[3] designed an automatic cleaning dust collection device for the inner torus of the lower row chain to solve the problem of inflexible action at the hinge of the chain plate. Li Lixun et al.[4] designed a set of pneumatic back-blowing tobacco leaf recovery units to solve the problem of the frequent shutdown of equipment caused by the blockage of filter rod holes. Ding Songfeng et al.[5] added a dust removal device to remove dust and recover cut tobacco to reduce cut tobacco loss. Xie Zhencheng and others[6] designed a rapid adjustment device for cutter head bearing clearance to reduce the frequency of cutter head channeling. Coke jump layer and others[7] improved the transmission system of the lower row chain of the wire cutter and reduced the frequency of the lower row chain sticking fault of the wire cutter. The above optimization and improvement improves equipment stability and reduces the consumption of cut tobacco, but does not solve the problems such as abnormal wear of copper bar chain. Therefore, the copper bar chain system of the TOBSPIN tobacco cutter improves comprehensively to reduce the wear speed of the copper bar chain and prolong its service life.

2. Problem Analysis

2.1. Working Principle and Structure

As shown in Figure 1, the compaction unit consists of a main compactor (Figure 3) with a copper bar chain (Figure 5) and a pre-compactor (Figure 4). A servo motor drives the compaction union with a reducer (Figure 1). The cutter door seat (Figure 2) presses the compaction unit from top to bottom to realize the compaction process. The tobacco leaves are thus compacted and transported to the cutter head through the cutter door[8].

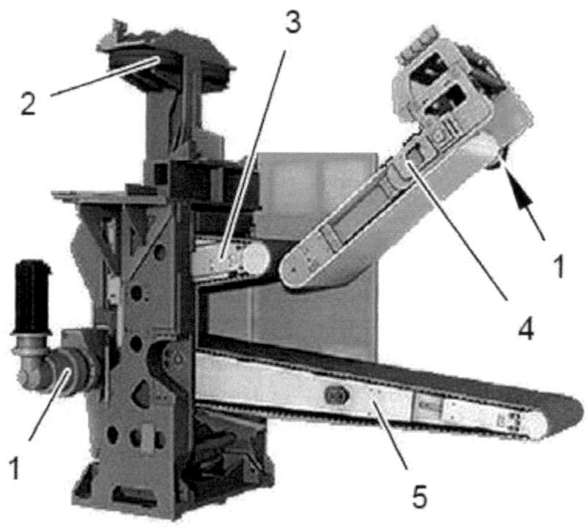

1. Reducer 2. Knife door seat 3. Main compactor 4. Pre-compactor 5. Lower chain
Figure 1. Structure diagram of compaction unit of TOBSPIN tobacco cutter

Figures 2 and 3 show that the copper bar chain is supported by the wear plate fixed on the frame. The motor drives the driving roller with teeth to drive the lower copper bar chain. The driven roller and its two ends are equipped with automatic centering devices of copper row chains, which can automatically correct the deviation of a single chain.

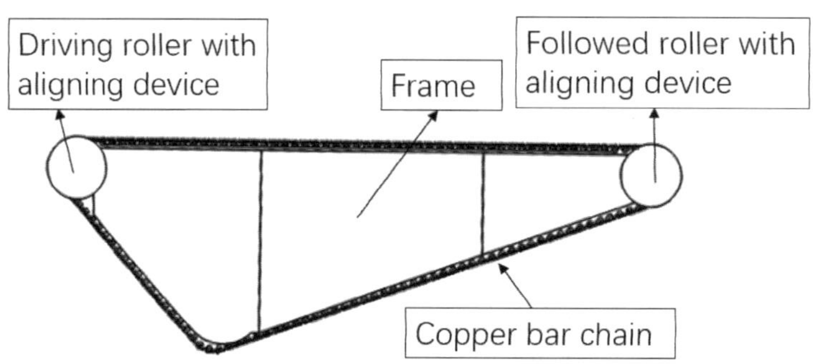

Figure 2. Schematic diagram of the copper row chain under the TOBSPIN tobacco cutter

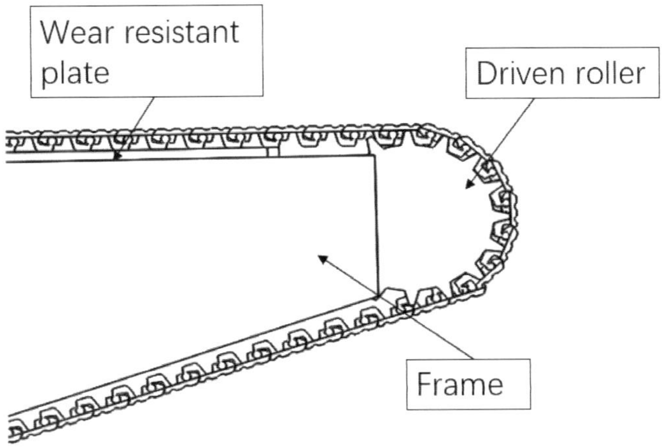

Figure 3. Local schematic diagram of copper row chain under TOBSPIN tobacco cutter

2.2. Problem Analysis
In daily maintenance, the TOBSPIN shredding machine has the following shortcomings.

According to TOBSPIN tobacco cutter use requirements[9], the gap between the copper exhaust chain and the frame is 1 mm under the new side. When the gap exceeds 5 mm, the copper exhaust chain must be replaced. During the operation of the TOBSPIN shredder, tobacco leaves enter into the copper bar chain through the gap between the copper bar chain and the wallboard, which not only causes waste of raw materials but also affects the regular operation of the copper bar chain. It also causes uneven left and right ends of the copper bar chain to collide with the frame, damaging the copper bar chain and bringing abnormal wear to the lower copper bar chain. It requires manual deep cleaning and maintenance of the shredder's copper bar chain at least once a week to ensure its normal operation. According to statistics, the wear rate of the copper bar chain is 2.62 mm/800h, and it needs to be replaced after about one year of use. Using it is more expensive.

3. Improvement Method
Given the existing problems of the TOBSPIN shredding machine, the equipment improvement of its copper chain wallboard and frame can improve the utilization rate of tobacco leaves and prolong the service life of the copper chain.

3.1 Improvement of Copper Chain Wallboard
In order to reduce the number of tobacco leaves and tobacco dust entering the gap between the copper chain and the frame, the wallboard of the copper chain is redesigned, and the lower copper chain area is changed from sealed to open.

As shown in Figure 4, the original whole wallboard 1 is replaced by a wallboard thicker than the chain board, and a bent material guide plate is added under the improved wallboard, with a gap of 20 mm in the vertical direction between the material guide plate and the improved wallboard. The elbow part of the material guide plate clings to the lower end of the wear-resistant plate, leading most of the cigarette dust and broken tobacco leaves to fall from the gap between the copper platoon chain and the frame out of the lower copper platoon chain area. The other end of the guide plate extends beyond the loose edge of the copper bar chain, preventing the fallen cigarette dust and broken tobacco leaves from falling into the annular inner surface of the copper bar chain and directly guiding the tobacco leaves to the frame. At the end of production daily, operators can clean the tobacco leaves on the rack.

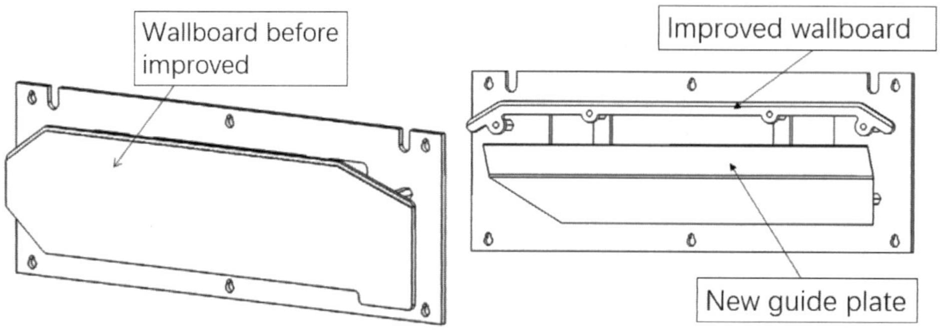

Figure. 4 The comparison of three-dimensional views of wallboard 1 before and after improvement

As shown in Figure 5, a whole wallboard 2 is changed to one-third of the original width. The wallboard cover is hollowed out, and plexiglass is installed in its place to observe the scale deposition of smoke dust on the annular inner surface of the copper platoon chain.

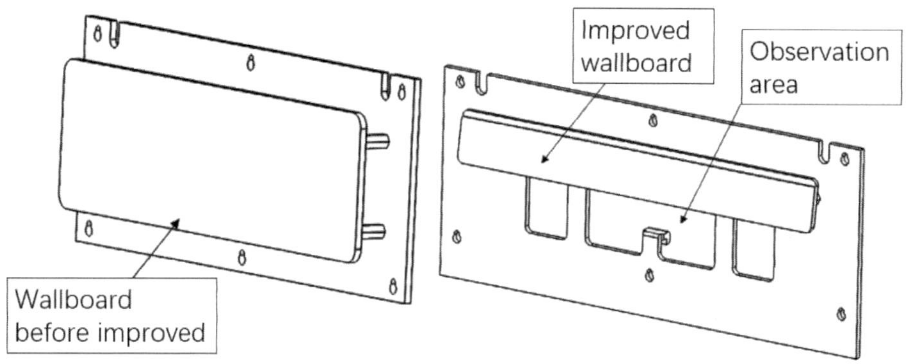

Figure 5. Comparison of three-dimensional views of wallboard 2 before and after improvement.

3.2. Improvement of Rack

By improving the copper bar chain's wallboard, most tobacco leaves falling between the copper bar chain and the rack are led out to the rack, but a few will fall between the copper bar chain and the wear-resistant plate. Therefore, the equipment of the rack is optimized again to improve the utilization rate of tobacco leaves.

3.2.1. Structural Design

Given the problem of the gap between a small number of minced tobacco and broken tobacco leaves falling into the copper row chain and the wear-resistant plate, the wear-resistant plate is redesigned, and the frame is improved accordingly. Moreover, the tight edge of the copper row chain is on the wear resistance plate, fixed to the first layer frame by the screws. The loose edge of the copper row chain is below the second layer and does not come into contact with the frame. Three final smoke collection ports were opened at the corresponding positions on the wear-resistant plate and the first layer rack, as shown in Figure 6.

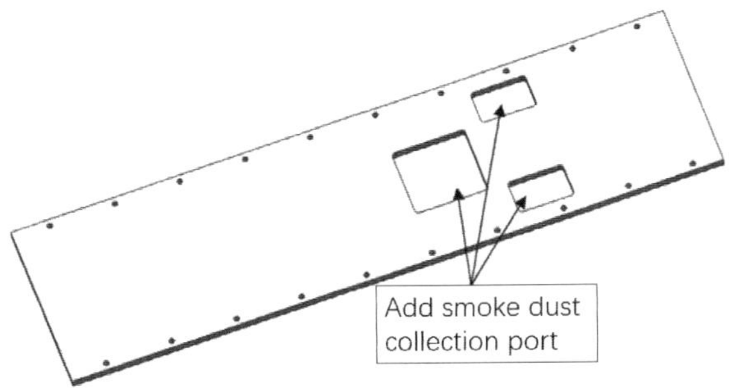

Figure 6. Schematic diagram of the improved wear-resistant plate below copper row chain

In order to ensure the strength of the rack and the effect of smoke exhaust, the smoke dust collection port is set at the middle and rear end of the lower copper platoon chain. When the copper platoon chain is running, unbonded broken tobacco leaves fall into the inner side of the tight edge of the copper platoon chain and end on the second rack. It is convenient for operators to use the vacuum cleaner to clean the second rack daily after production. The specific dimensions are shown in Figure 7.

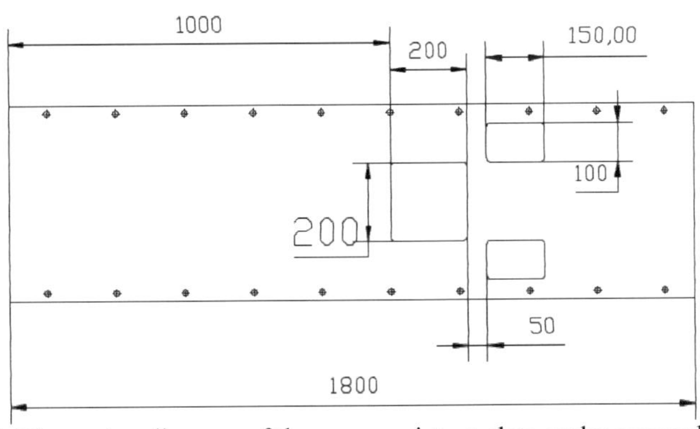

Figure 7. Dimension diagram of the wear-resistant plate under copper bar chain

3.2.2. *Analysis and Simulation*

According to the actual production situation, the frame is analyzed and simulated.

To simplify the analysis, the lower copper bar chain is idealized as a horizontal state (the actual force is less than the horizontal state). In the working state, the frame of the lower copper bar chain is under the pressure of the knife door seat of 32 kN; the gravity of the upper copper bar chain and the frame is 1,480×9.8=14,504 N; the gravity of the lower copper bar chain is 263.2×9.8=2,579 N; the gravity of the tobacco leaves above the copper bar chain is 300× 9.8 = 2,940 N.

Because the length of the upper copper bar chain in the horizontal direction is about 900 mm, the pressure of the knife door seat and the gravity of the upper copper bar chain and the rack are approximately uniformly distributed at 900 mm in front of the rack. The gravity of the lower copper bar chain and tobacco leaves are supported by the whole rack.

According to the above analysis, it can be calculated that the force acting on the front 900 mm frame is 3,200 N+14,504 N+0.5X (2,579 N+2,940 N) = 49,263.5 N, and the force acting on the rear 900 mm frame is 0.5X (2,579 N+2,940 N) = 2,759.5 N.

The frame's force analysis is carried out using HyperMesh software[10], and 20 mm at both sides of the frame is set as a fixed constraint, as shown in Figure 8.

Journal of Physics: Conference Series **2561** (2023) 012002 doi:10.1088/1742-6596/2561/1/012002

Figure 8. Setting diagram of frame constraints

The force directions of 49,263.5 N and 2,759.5 N applied to the left and right parts with the center line as the boundary are vertical and downward, which are uniformly distributed in the plane. The deformation distribution and stress distribution of the frame are shown in Figures 9 and 10

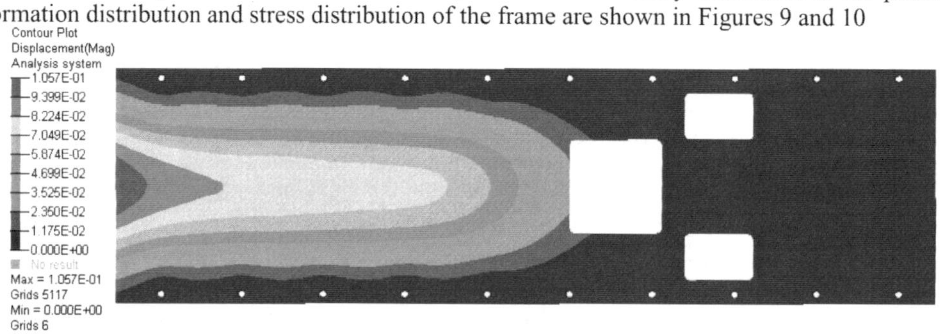

Figure 9. Deformation diagram of the frame

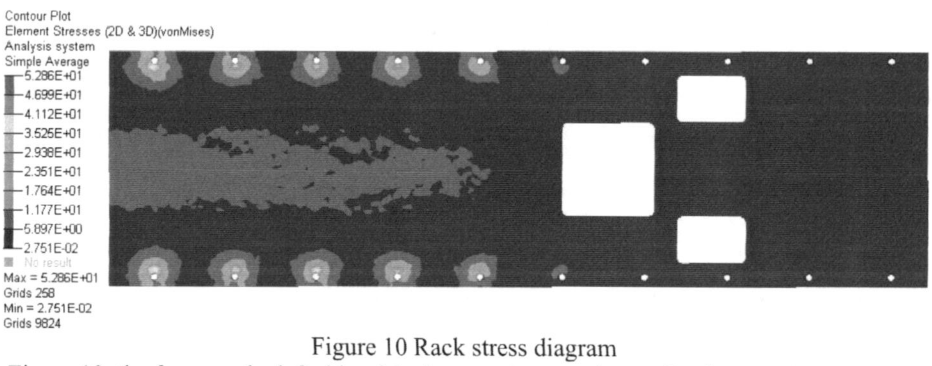

Figure 10 Rack stress diagram

In Figure 10, the force on the left side of the largest cigarette dust collection port is more significant. In order to better see the stress at the fresh hole, the unit on the left side of the big hole of the cigarette dust collection port is extracted separately, and the results are shown in Figures 11 and 12.

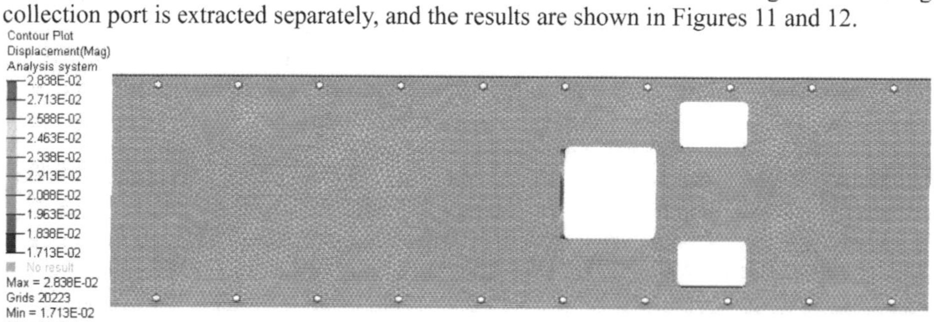

Figure 11. Deformation diagram of cigarette dust collection port

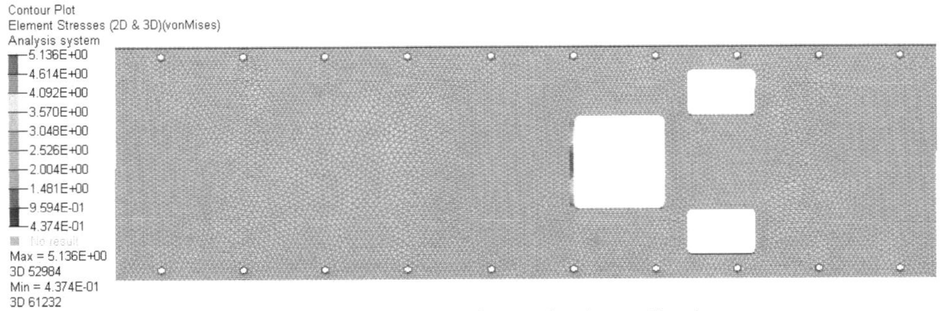

Figure 12. Stress diagram at the smoke dust collection port

It is clear that the middle deformation of the left edge of the cigarette dust collecting port is the largest, which is 0.028 mm, and the maximum stress is 5.1 MPa, far less than the minimum yield strength of cast iron of 190 MPa[11]. Therefore, the external force has little influence on the opening, and the stress and deformation at the opening are minimal, so plastic deformation will not occur.

3.3. *New Purging and Recovery Device*

In order to better clean up some tobacco leaves and shreds on the copper platoon chain, a 6-mm-diameter purging port is designed at the height parallel to the loose edge of the copper platoon chain, which synchronizes the copper platoon chain. A dust bag is installed opposite the purging port to collect and reuse tobacco leaves or shreds, thus improving the utilization rate of tobacco leaves and increasing certain economic benefits.

3.3.1. *Gas Circuit Diagram and Electronic Control Program Design*

Figures 13 and 14 show that the three-position five-way solenoid valve Y1 is energized and that the cylinder drives the upper copper bar chain and the upper knife door seat to press down. At the same time, the compressed air is delivered to the blowing pipe through check valve 2, pneumatic three-element 3, two-position three-way solenoid valve 4, and stop valve 5. The other end of the blowpipe is connected with the wallboard of the copper platoon chain through a threaded M6 air pipe quick connection. To prevent the air pipe's quick-connect thread from colliding with the uneven copper bar chain, the purging port is flush with the wallboard. The height of the purging port is the same as the loose edge of the copper platoon chain, so compressed air blows the loose edge inner ring of the copper platoon chain horizontally. A collection bag is installed on the frame opposite the nozzle, and the blown-out smoke powder is uniformly cleaned and recycled after daily production.

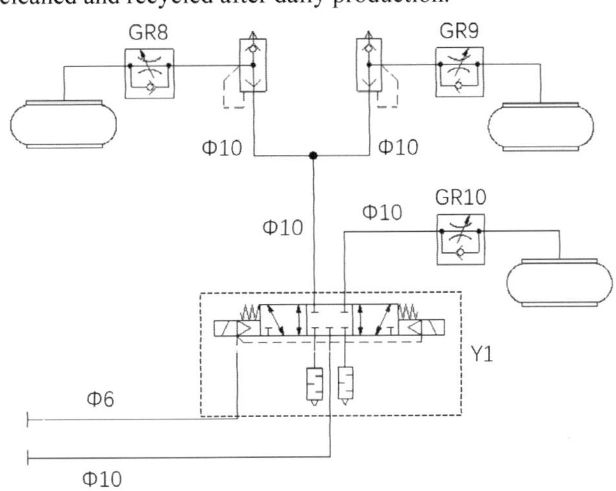

Figure 13. Control gas path diagram of compacting device

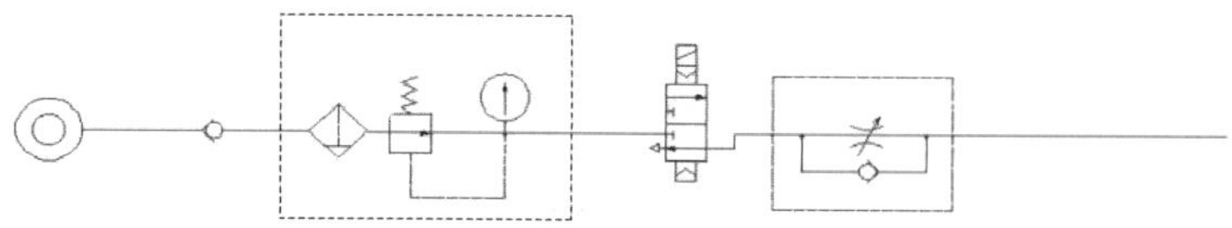

1 2 3 4 5

Figure 14. Control gas path diagram of purging device

The function of the synchronous design of the corresponding electronic control program is to interlock with the lower copper platoon chain. When the lower copper platoon chain runs, the purging device starts synchronously and horizontally blows the lower copper platoon chain. It also blows the cigarette dust into the ash collection bag. The flow chart of the blow-out air unit is shown in Figure 15.

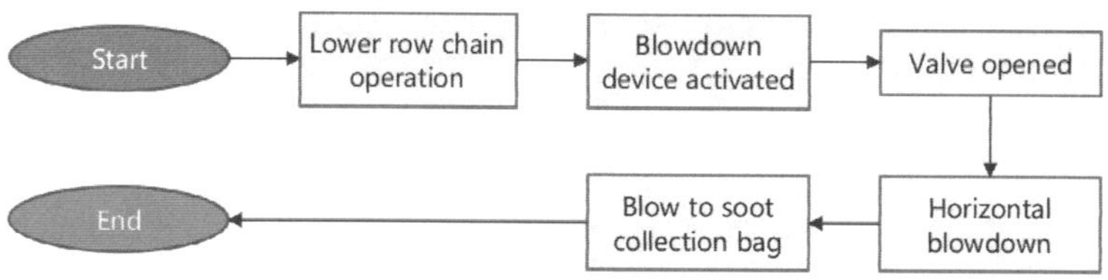

Figure 15. The flow chart of blow-out air unit

3.3.2. Selection of Compressed Air Pressure

In order to ensure the purging effect of compressed air, the compressed air pressure is determined in combination with the actual production situation. The formula for cutting width of a known cutter is as below:

$$b = \frac{V_A}{n_K \cdot N}$$

In the formula, b is the cutting width, mm; V_A represents the speed of the copper row chain, m / s; n_K stands for the speed of the knife roller, r / min; N is the number of blades.

In the actual production, the cutting width b is 1 mm, the knife roll speed $n_{\text{knife roll}}$ is 208 r/min, and the number of blades N is 8. $V_A = 208 \times 8 \times 0.001 = 1.664\ \text{m/min} = 0.028\ \text{m/s}$.

According to the actual measurement, the width of the chain groove part b_c is 5 mm. Then, the chain groove through the purge mouth time is $t = \frac{b_c}{V_A} = \frac{0.005}{0.028} = 0.179$ s.

If the width of the link b_S is 500 mm, the entire link is blown across to achieve the minimum speed of compressed air. $v_{min} = \frac{b_S}{t} = \frac{0.5}{0.179} = 2.8\ \text{m/s}$

It is known that the compressed air pressure of the silk-making workshop process p is 0.3 MPa. According to the air density table (shown in Figure 16), the corresponding air density of 25°C environment ρ is 3.5073 Kg / m^3, and the atmospheric pressure p_0 is 101 KP a[12]. According to the Bernoulli Eq$\frac{p}{\rho} + \frac{v^2}{2} + gh = const$[13], $\triangle p$ in this problem decreases to $\triangle p = p - p_0 = \rho\frac{v^2}{2}$, and an exhaust speed is provided: $v = \sqrt{\frac{2(p-p_0)}{\rho}} = 10.65\ m/s > v_{min}$.

Therefore, directly selecting the compressed air pressure of the process in the silk-making workshop can realize the purging of the copper bar chain's link gap.

Absolute pressure	Air temperature	Air density
Mpa	Celsius(℃)	Kg/m3
0.1	25	1.1691
0.2	25	2.3381
0.3	25	3.5073
0.4	25	4.6764
0.5	25	5.8455
0.6	25	7.0146

Figure 16. The density table of compressed air

3.3.3. Size Design of Dust Bag

The position of the purging port is 3 mm above the lowest point of the copper bar chain, as shown in Figure 17. To determine the purge opening's position and size, the software Fluent[14] is used to simulate the blowing, and the results are shown in Figure 18.

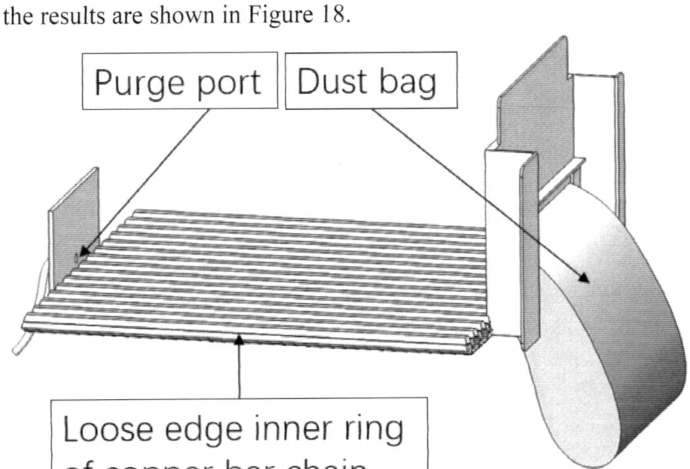

Figure 17. Schematic diagram of tobacco leaf purging and recycling device

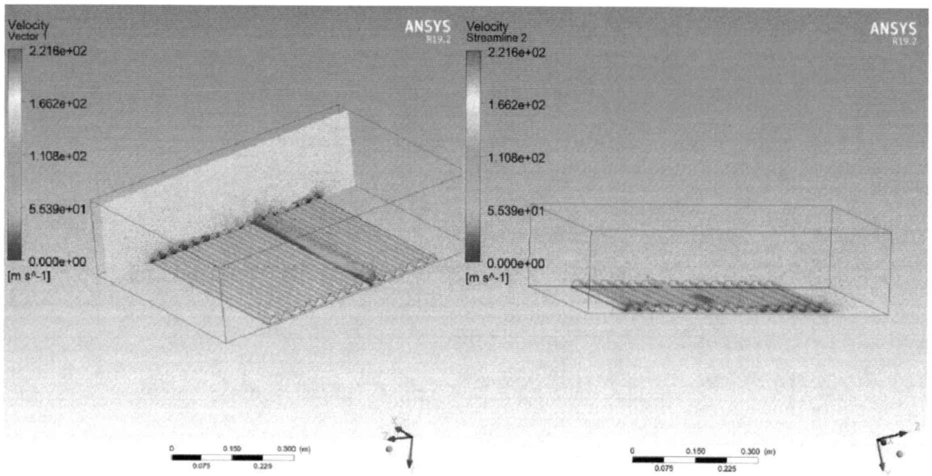

Figure 18. Simulation diagram of blow-out air recovery unit

From the figure, the compressed air blown out from the purge port is concentrated on the wall panel directly opposite the purge port, with a height of 0-50 mm and a width of 200 mm on the wall panel. Considering the size of the site and the effect of purging and dust collection, a dust bag should be set with an opening size of 100 mm×300 mm directly opposite the purging opening.

4. Application Effect

4.1. Experimental Design

Material: "Liqun" series tobacco leaves (provided by Ningbo Cigarette Factory of Zhejiang China Tobacco Industry Co., Ltd.)

Equipment: TOBSPIN shredding machine (HAUNI Company, Germany)

Methods: TOBSPIN shredding machine keeps producing in two shifts at a speed of 8,000 kg/h. Each line operates two devices in turn, and each device operates for about 8 hours daily according to the production schedule. According to equipment maintenance requirements, the copper bar chain needs cleaning and maintenance every five working days. The equipment running time, the gap between the copper bar chain, and the frame of two TOBSPIN shredders on the F line before and after the improvement are counted after the new copper bar chain is replaced for 20 weeks. The wear speed of the copper bar chain is calculated, and the average value is taken.

4.2. Data Analysis

As shown in Table 1, the wear speed of the improved copper bar chain is reduced from 2.62 mm/800 h to 0.85 mm/800 h, down by 67.6%. The service life of the improved copper bar chain is doubled from the original set of copper bar chains replaced every three years.

Table 1 Comparison of copper exhaust chain wear before and after the improvement

Machine number	Before the improvement①			After the improvement②		
	perfor mance period /h	Copper row chain with, rack clearance /mm	Copper drain chain wear speed /(mm/800 h)	Perfor mance period /h	Copper row chain with, rack clearance /mm	Copper drain chain wear speed /(mm/800 h)
F1	856	3.86	2.67	989	2.00	0.81
F2	808	3.61	2.58	918	2.04	0.91
F1	764	3.47	2.59	827	1.92	0.89

F2	783	3.58	2.64	814	1.83	0.82
Average	802.75	3.63	2.62	887	1.95	0.85

Note: ① The statistical time before improvement is March 6, 2020 and July 24, 2017; ② The improved statistical time is March 5, 2021 and July 23, 2018.

5. Conclusion

Through the comprehensive improvement of the copper bar chain system of TOBSPIN tobacco cutter, the lower copper bar chain wallboard is optimized. The material guide plate, the cigarette dust collection port of the lower copper bar chain, and the purging and recovery device are added, effectively reducing the wear on both sides of the copper bar chain and extending the service life of the copper bar chain. The TOBSPIN tobacco cutter used in Ningbo Cigarette Factory is tested. The results show that the wear speed of the improved copper bar chain decreased from 2.62 mm/800 h to 0.85 mm/800 h, down by 67.6%. The equipment maintenance cost drops, and the equipment's service life increases.

REFERENCES

[1] State Tobacco Monopoly Administration Cigarette Technology Specification [M] Beijing: China Light Industry Press, 2016.
[2] Chen Xuemei, Wang Yingli, Wenwu. "The control principle and the main characteristics for TOBSPIN cutter," Technology and Market, 2014,21 (2): 36-37 + 39.
[3] Zhang Zhigang. "Install inner ring surface cleaning device for the copper chain of TOBSPIN cutter," Shandong Industrial Technology, 2016 (19): 15.
[4] Li Lixun, Jiao Yang, Xu Renjie, et al. "Design of tobacco reclaiming unit with opposite-blow for TOBSPIN cutter," Tobacco Technology, 2017,50 (5): 94-99.
[5] Ding Songfeng, Liu Huaping. "Install dust-removing equipment for TOBSPIN cutter," Scientific Chinese, 2016 (17): 8.
[6] Xie Jingcheng, Tan Jiale, Liang Yue, et al. "Design of rapid adjustment device for bearing clearance of cutting blade carrier in TOBSPIN tobacco cutter," Tobacco Technology, 2022,55 (05): 108-112.
[7] Jiaocline, Xu Jianliao, Liao Zhongsheng, et al. "Modification of transmission system for bottom band in COMAS-A1 tobacco cutter," Tobacco Technology, 2020,53 (05): 99-104.
[8] Zhou Yusheng, Yang Bin, Li Hui, et al. "Effects of top band movement on consistence of cut tobacco width in rotary guillotine cutter," Tobacco Technology, 2010 (04): 21-23 + 29.
[9] HAUNI. TOBSPIN repair instructions [K].HWR.0580.001.01.zh.2013.
[10] Xiong Zhenbing, Luo Huixin "Preprocessing technology of FEA based on HyperMesh software," Drainage and irrigation machinery, 2006,24 (3): 4.
[11] An Huifen Metal Materials and Heat Treatment [M] Machinery Industry Press, 2014..
[12] Wang Yuzhong "Determine the historical evolution of standard atmospheric pressure," Physics teaching discussion: Chinese Teaching Edition, 2001,19 (10): 1.
[13] Chen Yanli. "Analysis of the Principle of Bernoulli Equation and Application," Journal of Luohe Vocational and Technical College, 2012,11 (2): 3.
[14] Li Yong, Liu Zhiyou, An Yiran "A brief introduction to Fluent— a general purpose CFD code," Hydrodynamic Research and Progress: Series A, 2001,16 (2): 5.

Design and application of copper chain removal device for TOBSPIN tobacco cutter

Li Xu, Yubin Fang, Min Xu, Kelei Sun*, Bin Lin, Jun Zhu, Yujie Wu

Ningbo Cigarette Factory, China Tobacco Zhejiang Industry Co., Ltd., Ningbo 315040, China

* Corresponding author: 417612486@qq.com

ABSTRACT In order to solve the difficulty in disassembling the copper chain of the TOBSPIN tobacco cutter, an automatic disassembly device was developed based on Solidworks mechanical simulation design software. Firstly, construct the overall structural model of the device by Solidworks. Then determining the feasibility of the equipment operation by the force analysis of the system. Lastly, the device can be automatically started and stopped by controlling PLC program. The results show that it takes 41.24 hours for manual disassembly per machine, and 11.18 hours to apply the automatic device, which reduces the working time by 72.9%. The device can provide technical support for improving the disassembly efficiency of the equipment and reducing operating intensity.

1. INTRODUCTION

The cutting process is an important part of cigarette production. The tobacco cutter is used to cut tobacco leaves into pipe tobacco, therefore, its performance directly affects the quality of the pipe tobacco [1]. TOBSPIN tobacco cutter is a novel rotary cutting equipment produced by German company HAUNI which has the advantages of a high degree of automation, easy operation and low noise [2]. Meanwhile, it can significantly reduce tobacco leaf loss during the process and ensure quality consistency. The copper chain is a pivotal part of the feeding structure for cutting equipment. During production, the copper chain contacts the material directly, and a large squeezing force is applied to the material. Therefore, scraps and smoke oil in the material are easy to adhere to the surface of the chain and enter the gap of connecting shaft [3]. If the copper chain is too dirty, the tobacco leaves will slip during the conveying process, resulting in uneven width. Therefore, the copper chain needs to be disassembled and maintained regularly to ensure normal performance.

At present, various types of research have been investigated on TOBSPIN tobacco cutters. Zhao Chenguang et al. [4] designed a cleaning machine to realize the automatic cleaning of copper chains. Dai Yuxin [5] designed a set of water mist lubrication application devices to solve the serious wear problem of copper chains and wear plates. However, the disassembly problem of the copper chain has not been solved. It is proposed to design a copper chain disassembly device for TOBSPIN tobacco cutter, improving the disassembly efficiency and reducing the operator's labor intensity.

Compared with other traditional cutters, there is a fundamental difference in the connection method of the copper chain for TOBSPIN tobacco cutter. That is, the original bulkhead plug is changed to a single-forming connection [6]. This connection method reduces the gap between the copper chains, and the tobacco leaves on the surface of the copper chain will not clip the gap, which effectively keeps the inside of the copper chain clean, and the transmission is more reliable.

However, in the actual production process, the copper chain is spliced by pieces of movable plates.

A lot of tobacco leaves and tobacco dust always enter into the side gap of the copper chain during running, which makes a bulge form in the copper chain. As the tobacco leaves are gradually squeezed into the copper chain, it will lead to a further increase in the deformation angle of the copper chain which affects the performance and must be replaced and cleaned. The structure of the TOBSPIN tobacco cutter copper chain is shown in Fig 1.

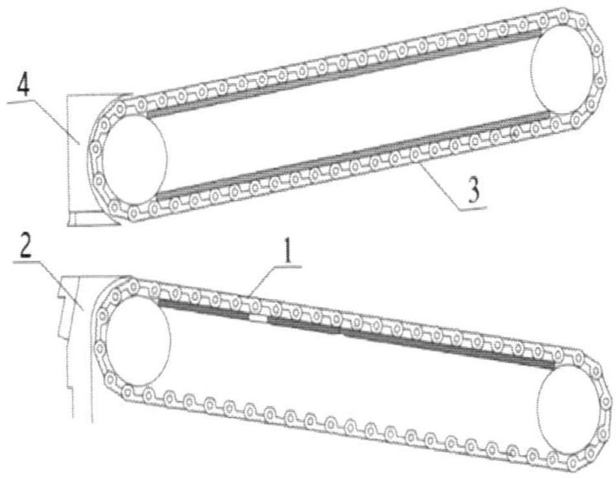

Fig.1 Schematic diagram of copper chain
(1- Lower copper chain, 2-Lower knife door, 3-Upper copper chain, 4-Upper knife door)

According to the requirement of the equipment maintenance manual [7], combined with the use of the equipment, the replacement and cleaning cycle of the copper chain is about 4 months. While the operation must be disassembled one by one. However, the copper chains rotate with each other, which makes disassembly very difficult and time-consuming and labor-intensive.

2. MEASURES

The mechanical structure of the copper chain disassembly device of TOBSPIN tobacco cutter is mainly composed of a supporting frame platform, motion triplet, screw and other components, as shown in Fig 2. The support plate 3 is adjustable, plates 4 and 5 are welded with screw 6, plate 4 is welded with the protection fence 8, and plate 5 is fixed on the movable support platform 10.

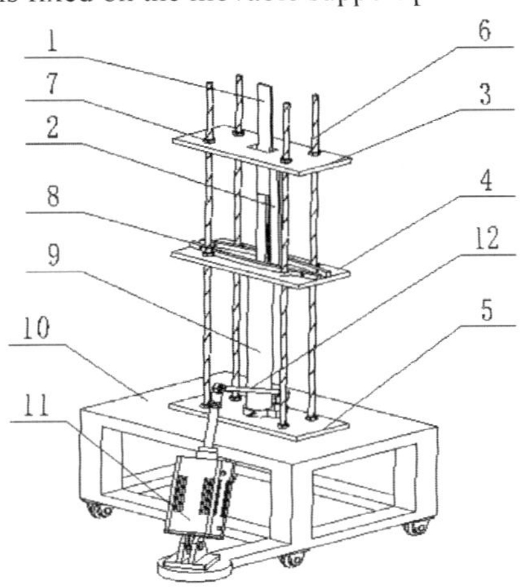

Fig. 2 Mechanical structure of copper chain automatic disassembly device
(1, 2-copper chain, 3, 4, 5-support plate, 6-screw, 7-nut, 8-protection rod, 9-jack, 10-support platform, 11-cylinder, 12-main push rod)

When disassembly the copper chain of the wire cutter, first of all, it is necessary to install the copper chain 1 and 2 between the support splints 3 and 4, and adjust the height of the support plate through nut

7 on the screw 6 to fix it, protect fence 8 can also prevent the displacement of the copper chain 1 and 2. Then we connect the compressed air end to the interface of cylinder 11, and dri0076ing jack 9 circularly by swinging the main push rod 12. Finally, copper chain 1 moves up in a straight line and is pushed out from the small hole of support plate 3.

2.1 Hydraulic jack

The hydraulic jack is a kind of jack that uses a hydraulic cylinder as a rigid lifting member [8]. It has some advantages such as a compact structure, stable operation and self-locking function. The hydraulic system consists of power components (hydraulic pumps), actuators (piston cylinders), control components (check valve and globe valve), auxiliary components (fuel tank and tubing, etc.), and a working medium (hydraulic oil) [9]. The structure of the hydraulic jack is shown in Fig 3.

When the handle is lifted up, it drives the piston upward and checks valve 1 is closed. In this way, the piston volume forms a vacuum, and then the tank liquid flows into the piston cylinder through check valve 2 under the atmospheric pressure. On the contrary, when the handle is pressed down, the piston moves down and closes check valve 2. The oil in the piston cylinder enters the plunger cylinder to lift the piston and separate the copper chain. If we need the plunger to stop, just stop the movement of the pressing rod and close check valve 1, the plunger is self-locking, and when the plunger is needed to return down, open the globe valve and reset the plunger [10].

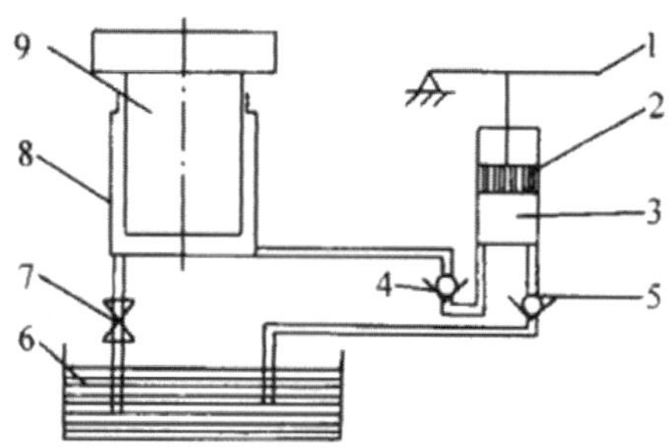

Fig.3 Schematic diagram of the hydraulic jack
(1-handle, 2-piston, 3-piston cylinder, 4, 5-check valve 1, 2, 6-fuel tank, 7-globe valve, 8-plunger cylinder, 9-plunger)

2.2 Cylinder

The cylinder consists of cylinder tube 1, front and rear cylinder head 2, piston rod 3, seals and fasteners [11]. The sectional view of the cylinder is shown in Fig 4. The pneumatic drive has advantages like high flexibility, low noise, no oil pollution, no external mechanical parts such as gears, relatively simple structure. Meanwhile, it can provide sufficient output force [12]. According to the mechanical input power requirements of the hydraulic jack, a fixed double-acting single-piston rod cylinder is selected for vertical installation [13].

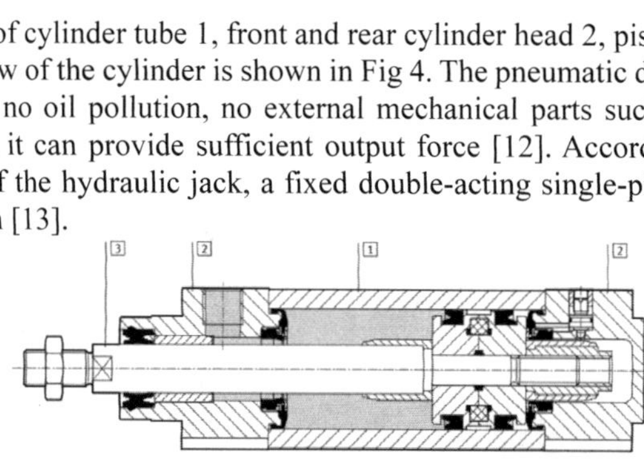

Fig.4 Sectional view of a cylinder
(1-cylinder tube, 2-front and rear cylinder heads, 3-piston rod)

Assuming that the maximum pressure of the cylinder gas source is p, the load rate of the cylinder (η=0.5) is selected according to the running conditions of the system, and the cylinder bore and rod diameter are D1 and D2, respectively. The maximum tension of the cylinder can be known from the gas pressure:

$$F_{\max}=\frac{\pi}{4}\times\left(D_1^{\ 2}-D_2^{\ 2}\right)\times p\times\eta \tag{1}$$

Where: Fmax is the maximum theoretical output force at the end of the piston rod (N); P is nominal pressure (MPa); D1 is the inner diameter of the cylinder (mm); D2 is the diameter of the push rod (mm).

2.3 Analysis of connecting rod force

Select a 5 tons hydraulic jack. The handle of the hydraulic jack is the connecting rod. The diameter of the large piston (d1) is 40 mm, and the small piston (d2) is 36 mm. According to Pascal's principle:

$$F1/F2=S1/S2 \tag{2}$$

Where F1 is the axial force to disassemble the copper chain, F2 is the axial force applied to the small piston by the connecting rod, S1 is the cross-sectional area of the large piston, and S2 is the cross-sectional area of the small piston.

Calculated by formula (1), $F2 = F1 * S2/S1$. According to the law of force interaction, the reaction force of the small piston on the handle of the hydraulic jack is F2. Since the hydraulic jack and cylinder are both vertical, it is easy to choose a slider to connect the output end (cylinder piston) of the cylinder and the input end (connecting rod) of the hydraulic jack based on the knowledge of mechanical principles. Assuming that when the hydraulic jack pries the copper chain to make it move relatively, the connecting rod just moves downwards at the highest position. At this time, the maximum force analysis of the required cylinder pulling force F3 is shown in Fig 5.

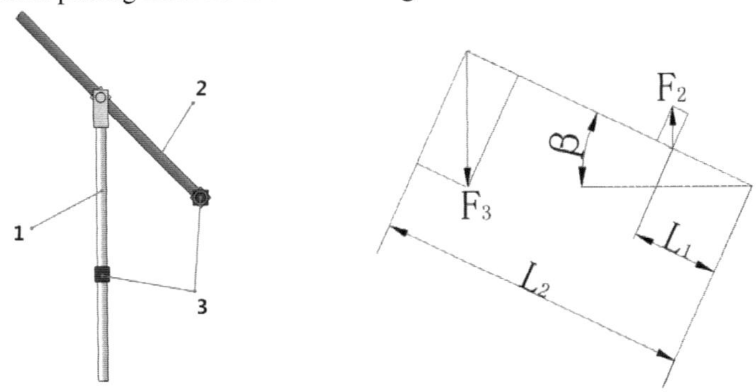

Fig.5 The three-dimensional model and plane force analysis diagram of the connecting rod (1-cylinder shaft, 2-hydraulic jack handle, 3-frame)

It can be seen from the torque balance condition:

$$F_3\cos\beta\cdot l_2=F_2\cos\beta\cdot l_1 \tag{3}$$

Where: l1 is the distance from the pivot point of the connecting rod to the action point of the small piston, l2 is the distance from the pivot point of the connecting rod to the action point of the cylinder piston rod, and β is the angle between the connecting rod and the horizontal plane.

According to Formula (2), F3=F2l1/l2. If F3>Fmax is specified, then the system can complete the disassembly of the copper chain.

2.4 ANSYS Simulation

Since the hydraulic cylinder is a standard part and has been verified to meet the design requirements, ANSYS simulation is performed only for the frame of the copper row chain removal device. According to the actual disassembly of the copper row chain, the maximum force required is 4900 N, which is applied to the corresponding part of the frame and the corresponding constraints are applied to the frame, as shown in Fig.6.

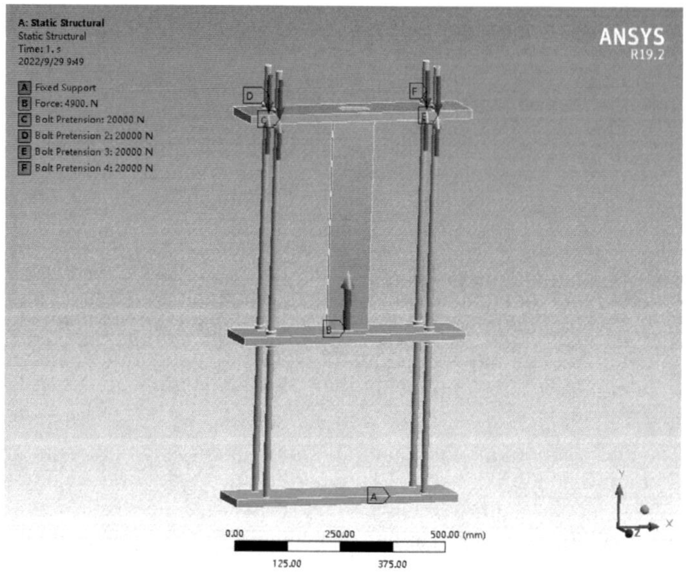

Fig.6 Diagram of the frame after applying constraints and forces

As shown in Fig.7, the maximum pressure on the frame is 158.7 MPa, and the frame is made of 45# steel with a yield strength (≥355 MPa). Therefore, the frame meets the design requirements of the device.

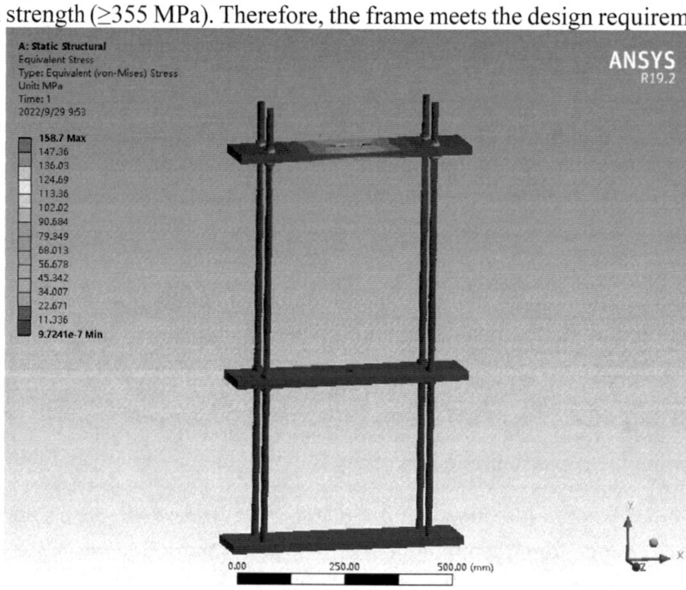

Fig.7 Results of the ANSYS simulation

2.5 Electronic control system

At present, the technology in the field of automation control is becoming more and more mature, and the control functions of equipment are closely related to benefits. Currently, most of them adopt electrical control [14]. In order to realize the close connection between these control components, the PLC control system is utilized to complete.

The electrical control part uses Siemens PLC programming software STEP7 to draw the ladder diagram. The automatic disassembly electrical control system designed with a programmable controller has a simple circuit, high reliability and strong function, and can realize the automatic disassembly of the copper chain [15].

According to the requirements of the control system in this research, the ladder diagram is shown in Fig 8. The working principle: when the system is powered on, the DC voltage of 24 V is supplied to the sensor, and the electrical signal collected by the sensor is transmitted to the PLC. After the PLC executes the user program and outputs a switch control signal to control the action of the solenoid valve, and then control the movement track of the cylinder and jack to complete the jack lifting out the copper chain. Finally, the jack will return to its original position.

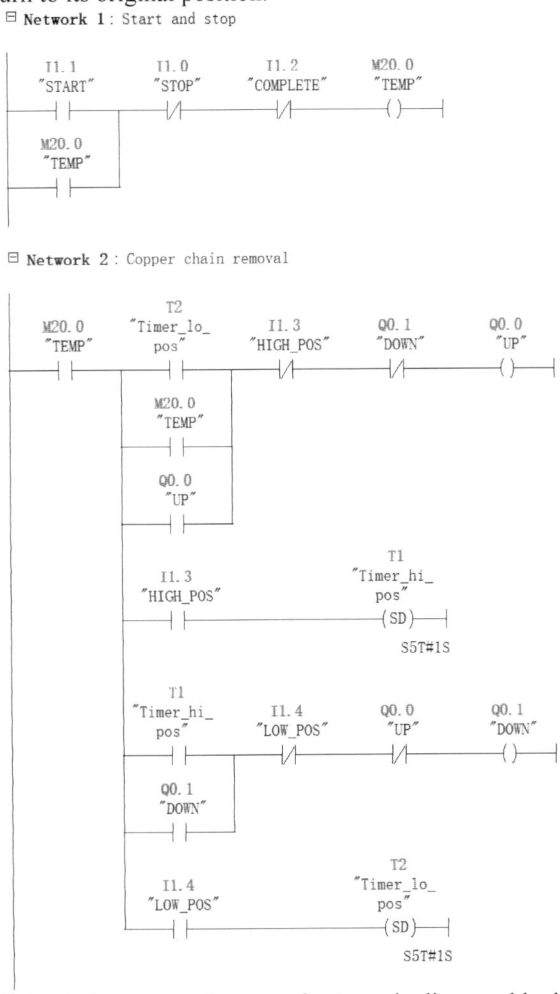

Fig.8 Control program diagram of automatic disassembly device

3. APPLICATION EFFECT

3.1 Experimental design

Equipment: A pair of copper chains for two TOBSPIN tobacco cutters (HAUNI company), the disassembly device (Ningbo Cigarette Factory)

Method: Select 1 pair of TOBSPIN tobacco cutter copper chains, use two methods of manual and automatic disassembly respectively, repeat 5 times, record the time and take the average value.

3.2 Data analysis

It can be seen from Table 1 that the average time of manual disassembly of the copper chain is 41.2 h/pair, and automatic disassembly is 11.18 h/pair. After using the device, disassembling a pair of copper chains saves an average of 30.06 h (72.9%). In addition, after the improvement, the equipment runs stably, which effectively improves the disassembly efficiency of the copper chain and reduces the operation intensity.

Table 1 Comparison of the disassembly time of the copper chain for TOBSPIN tobacco cutter (h)

No.	test time before the improvement①	test time after the improvement②	time difference
1	42.2	12.3	29.9
2	41.5	11.5	30.0
3	40.4	11.7	28.7
4	40.9	10.5	30.4
5	41.2	9.9	31.3
average	41.24	11.18	30.06

Notes: ①The test time before the improvement is from March to April 2020, and the copper chain is disassembled by the original method; ②The test time after the improvement is from March to April 2021, and the copper chain is disassembled by utilizing the disassembly device.

4. CONCLUSION

We design a disassembly device for TOBSPIN tobacco cutter copper chain, which replaces the original manual disassembly method and realizes the automatic disassembly. The device is driven by the electronic control system to control the solenoid valve to drive the movement of the cylinder. Then the jack lifts the copper chain to separate by the action of the cylinder.

Taking the copper chain of Ningbo Cigarette Factory as the test object, the results show that the average time of manual disassembly is 41.2 h/pair, and automatic disassembly is 11.18 h/pair. After improvement, a pair of copper chains save 30.06 h on average, and dismantling efficiency is greatly improved.

REFERENCES

[1] YANG B, Y, WEN Liangkui, CHEN Daoxiang. "Application of high speed and low material level technology in TOBSPIN cutter" [J]. Engineering Technology and Application, (09): 98-99 (2018).
[2] Chen Xuemei, Wang Yingli, Wen Wu. "The control principle and the main characteristics for TOBSPIN cutter" [J] Technology and Market, 21(2):3 (2014).
[3] JIANG Zhuofang, LI Lixun, WU Heng, etc. "Development of a dismantling tool for knife feeding motor in TOBSPIN wire cutter" [J]. Tobacco science and technology, 51 (08): 105-108 (2018).
[4] ZHAO Chenguang, WANG Xiaogang. "Development of Machine Cleaning for Compression Bands in Tobacco Cutter" [J]. Tobacco science and technology, (11): 17 +28 (2005).
[5] DAI Yuxin. "Design of water spraying device for compression bands lubrication in COMAS-A1 tobacco cutter" [J]. Tobacco Science & Technology, 49(5): 100-104 (2016).
[6] LU Zhongfu. "Discussion on automatic lubrication device for conveying copper bar chain of SQ3X wire cutter" [J]. China high tech enterprise, (26): 86-88 (2014).
[7] HAUNI. "TOBSPIN maintenance manual" [K]. HWR. 0580. 001. 01. zh. (2013).
[8] HE Xuehong, LU Wenjia, YIN Qizhen, etc. "Installing position optimization of hydraulic cylinder of scissors lift platform" [J]. Journal of Mechanical & Electrical engineering, 33 (10): 1159-1163 (2016).
[9] CHENG Sanhong, SHAO Guojian, XU Yunkui. "The Research and Application of Strengthening Test for Hydraulic Cylinder" [J]. Journal of Mechanical & Electrical engineering, (11): 16-19 (2004).

[10] YAN Taoping, WANG Jiansheng. "Design Improvement in the Structure and Oil Circuits of Hydraulic Jack" [J]. Construction machinery and equipment, (03): 38-39 (2002).
[11] SUN Jianhui, NI Xuguang, YUAN Qiaoling, etc. "Structure design and simulation of frictionless cylinder-piston based on Fluent" [J]. Journal of Mechanical & Electrical engineering, 30 (08): 933-936 (2013).
[12] ZHU Xiaoping, LIU Zhentao, YU Xiaoli. "Analysis of thermal-mechanical coupling strength and fatigue life of cylinder head" [J]. Journal of Mechanical & Electrical engineering, 28(10):1176-1179+1198 (2011).
[13] ZHAO Mengyun, CHENG Guijian, LI Qiang, etc. "Movement law of valve plate and simulation of cylinder state of reciprocating compressor under variable load condition" [J/OL]. Journal of Mechanical & Electrical engineering, (05): 480-484 (2019).
[14] ZHOU Li. "The Design of Frequency Conversioned and Isobarically Fuzzy Control Water Supplying System Based on PLC" [J]. Journal of Mechanical & Electrical engineering, (03): 14-17 (2005).
[15] SUN Kangling, YANG Zhaowei, ZHANG Ye. "Control system design of automatic door based on PLC" [J]. Journal of Mechanical & Electrical engineering, 27(11): 123-126 (2010).

Resonance and sensing characteristics of quartz tuning fork for force sensor

Yitian Ding[1], Jiahui Wang[2,*]

[1] Guangzhou Foreign Language School, Fenghuang Avenue 102, Guangzhou 511400, Guangdong, China.

[2] School of Physics, Sun Yat-Sen University, Guangzhou 510275, Guangdong, China

[*] Corresponding author: wangjh@mail.sysu.edu.cn

Abstract. Quartz tuning fork (QTF) is often utilised for the design and manufacture of high-performance sensors, such as the force sensor for atomic force microscopy (AFM). In this study, the mechanical resonance and sensing characteristics of QTF with an additional mass on its prongs are studied by theoretical analysis and experiments. The results show that the position and size of additional mass on the QTF prong have an obvious influence on its resonance frequency and quality factor (Q-factor). The greater the mass of the additional component or the closer to the free end of the QTF prong is, the lower its resonance frequency and quality factor will become. Reducing the additional effective mass can significantly improve its quality factor, thus improving the sensing performance of the QTF sensor. This study can provide a quantitative reference for the design and optimisation of the QTF force sensor for AFM.

1. Introduction

A quartz tuning fork (QTF) is the core component of a quartz crystal oscillator, which is an electronic component commonly used in electronic watches [1]. QTF has the characteristics of high-frequency stability and a high-quality factor (Q-factor) of its mechanical resonance, which can be utilised for the design and manufacture of various high-precision sensors. The temperature sensor made of QTF has been used in the field of man-made spaceflight [2]. QTF sensors can also be utilised to make high-precision and rapid humidity detection in the industry [3], to measure fluid density and viscosity [4], to make a photoacoustic gas sensor [5] or to use as a force sensor [6,7].

QTF has been widely used as a force sensor in atomic force microscopy (AFM) [7-17]. AFM usually utilises a micro-cantilever probe for laser detecting in its measurement and scanning imaging. In recent years, the force sensor based on QTF has been widely used in AFM, such as A-probe [9], which is used in the atmospheric environment.

The force sensor based on QTF in AFM has been studied a lot. Using classical physical theory, models and formulas, physical parameters such as resonance frequency and Q-factor of mechanical vibration of QTF can be analysed and calculated [10-15]. Ideally, the two prongs of a QTF are perfectly symmetrical. Nevertheless, when making a force sensor with QTF, it is often necessary to attach additional components, such as the sensor probe, to the prong of the QTF. These additional components will undoubtedly destroy the original symmetry of QTF, thus affecting the resonance frequency and Q-factor. They have a significant negative impact on sensing performance. Previous researches show that by reducing the additional mass [11] on the QTF prongs or rebalancing these additional mass [11,16,17] on the QTF prongs, the symmetry of the QTF can be restored to some extent, thereby improving its Q-factor and sensing characteristics.

Nevertheless, there are still some obstacles when the QTF force sensor for AFM is used in the atmospheric environment. On the one hand, due to damp air, the Q-factor of QTF in the atmospheric environment will be significantly reduced. As the minimum detectable force is inversely proportional to the square root of Q, a higher Q-factor is indispensable for the high force sensitivity of a force sensor [9]. On the other hand, when making a sensor with a QTF, it is usually necessary to attach a sensor component to the QTF prongs. It will inevitably affect the resonance characteristics of the QTF, such as lowering its Q-factor and resonance frequency.

The impact of additional mass and mass rebalance on QTF force sensors has been studied previously [11,16,17]. However, the influence of additional mass and its position on the mechanical resonance and sensing characteristics of QTF has not been systematically studied. There are few quantitative reference data for the design of the QTF sensor yet. The starting point of this study is to demonstrate the influence of additional mass upon the QTF prongs on the resonance sensing characteristics quantitatively to help improve the sensing performance of the QTF force sensor typically used in AFM.

2. Experimental principle, material and methods

2.1 Experimental material

In our experiments, we used QTFs with a resonance frequency of 32.768 kHz (YT-38 model, Shenzhen Yangxing Technology Co., Ltd, China). The photos of QTF are shown in Figure 1. The QTF has two sets of symmetrically distributed electrodes. When the electric excitation is used by applying a sine wave voltage on the electrode, the mechanical vibration of the QTF prongs in anti-phase mode can be excited. When the QTF is subjected to external force or additional mass, its mechanical resonance frequency will change. Electric excitation and electric detection can be adopted by the QTF sensor, leading to the advantages of simple structure and high accuracy. The measured dimensions of the employed QTF in our study are shown in Table 1.

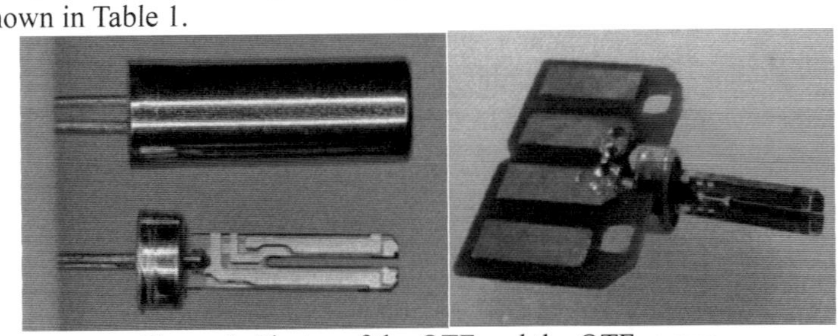

Figure 1. Photos of the QTF and the QTF sensor.
(The left photo shows a QTF before and after shelling, and the right shows a QTF force sensor.)

Table 1. Dimensions of the QTF (Unit: mm)

	The QTF	Prongs of QTF
Length	6.06	3.76
Width	1.45	0.59
Thickness	0.32	0.32

2.2 Theoretical background and estimation of QTF parameters

In the spring model, let the spring constant be k, and the mass m, then the eigenfrequency f of spring mechanical vibration is [11]:

$$f=\frac{1}{2\pi}\sqrt{\frac{k}{m}} \quad (1)$$

For a single micro-cantilever, k is the elastic factor in the tangent direction, and m is the effective mass.

In a simplified model, the calculation of k values can be derived from the physical dimensions of the QTF and the physical parameters of quartz material:

$$k = \frac{1}{4} E w (\frac{t}{L})^3 \tag{2}$$

Among them, E is Young's modulus of quartz material, $w, t, and\ L$ are the width, thickness and length of the QTF prong, respectively. In a simplified model, the effective mass m_{eff} can be calculated by the following formula [12]:

$$m_{eff} = \alpha w t L \rho \tag{3}$$

where ρ is the density of the quartz material, α is the ratio between effective mass m_{eff} and total mass m. For rectangular micro-cantilevers, α can be $\frac{1}{4}$ most of the time.

k and m are input into the above Formula (1), the formula of the eigenfrequency f becomes [14]:

$$f = \frac{1}{4\pi} \frac{t}{L^2} \sqrt{\frac{E}{\alpha \rho}} \tag{4}$$

This formula is usually used to calculate the eigenfrequency of a single prong or a cantilever.

Dimensions of the QTF used in this experiment are shown in Table 1. By substituting Young's modulus ($E = 7.87 \times 10^{10} Pa$) and density ($\rho = 2650\ kg \cdot m^{-3}$) [15] and the above data into Formula (4), it gives: $f = 36.378\ kHz$. The calculated resonance frequency f is similar to that provided by the manufacturer, which is $32.768\ kHz$, giving a relative error of 11%. The main source of this error is that when calculating, the effective mass above the value of $\frac{1}{4}$ is used for α. When α is adjusted to 0.308, it gives $f = 32.774\ kHz$, which is closer to the parameter supplied by the factory.

The vibration frequency and Q-factor of the resonance mode of QTF will change when the quality of the two prongs differs in Δm [11].

$$f_0' = \frac{1}{2\pi} \sqrt{\frac{k}{m+\Delta m}} = f_0 \sqrt{\frac{m}{m+\Delta m}} \tag{5}$$

$$Q_0' = \sqrt{\frac{m}{m+\Delta m}} Q_0 \tag{6}$$

It can be seen that the Q-factor can be improved by improving the mass balance between the two prongs of the QTF [11,16,17].

When making a force sensor with QTF, the minimum force gradient $\delta F'_{min}$ that can be detected through the mechanical vibration of the QTF has the following relationship [9].

$$\delta F'_{min} \propto f_0^{-\frac{1}{2}} Q^{-\frac{1}{2}} \tag{7}$$

Thus, the minimum force $\delta F'_{min}$ that can be detected is inversely proportional to the square root of the Q-factor Q. That is, the larger the Q is, the smaller the minimum force $\delta F'_{min}$ that can be detected, so the higher sensitivity of the force sensor and the better performance of the sensor. For an atmospheric force sensor based on QTF, when a probe is attached to its prong, the Q-value will be significantly reduced. For example, when the Q-value decreases from above 10,000 to below 1,000, the minimum detectable force $\delta F'_{min}$ will significantly reduce accordingly. Therefore, improving the Q-value of the QTF force sensor is an effective way to improve its force sensitivity. Therefore, the Q-factor of QTF can be employed as the key factor of the sensor sensitivity.

2.3 Bonding micro-balls onto the QTF

2.3.1 Experimental material and methods
Using the unshelled QTF as the material, a micro-ball with a certain quality (simulating the additional mass of the sensing probe attached to the QTF prongs) was pasted to the QTF prongs as the experimental sample. The experimental method is similar to that described in [17].

The micro ball used in this experiment is a lead-tin ball with a diameter of $0.2 - 0.6\ mm$ (tin 63%, lead 37%). The glue used for bonding is liquid glue that dries slowly in 100 seconds (501 type slow drying glue, Guangzhou Zhuolide Co., Ltd.).

2.3.2 Adhesion of micro balls onto the QTF prong
The experimental steps for bonding the micro ball onto the QTF prong are shown in Figure 2. The QTF welded on the bracket is clamped and fixed on a four-dimensional translation platform. To bond the micro ball onto the QTF prong, mainly include three steps:

Step 1: The glue is dispensed. As shown in Figure 2 (left), a sharp needle head is fixed on the three-dimensional translation platform. A small amount of slow-drying glue is absorbed. The needle head is moved onto the QTF prong by adjusting the platform so that a layer of glue can be pasted onto the appropriate position of the QTF prong.

Step 2: The micro ball is selected. As shown in the middle of Figure 2, the micro balls are first poured into a small container and put under an optical microscope. Then, a micro ball is picked using tweezers and clamped while the clamping position should be as close as possible to the front end of the tweezers.

Step 3: The ball is placed. As shown in the right of Figure 2, the optical microscope is focused on the glued position. Then the tweezers are moved to hold the micro ball. We then move the micro ball slowly. After the ball contacts with the QTF prong surface, the glue area is moved back and forth several times to make more glue stick to the ball, then the tweezers are removed, and we wait for the glue to solidify.

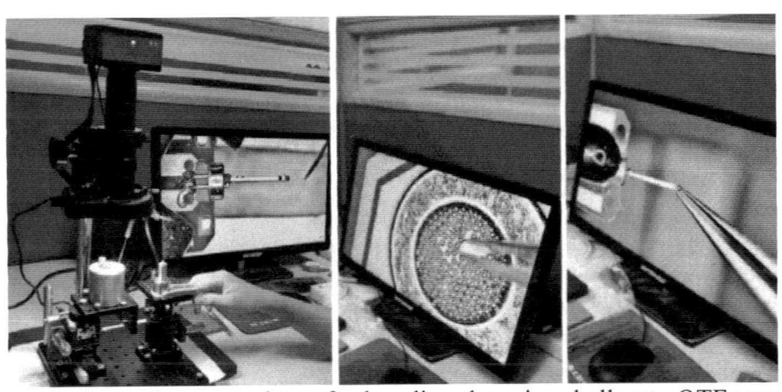

Figure 2. Operating procedures for bonding the micro ball onto QTF prongs.
(Left: The glue is dispensed; Middle: The micro ball is selected; Right: The ball is placed onto the selected position)

2.4 Measuring apparatus and methods
The resonance frequency and Q-factor of QTF can be acquired by measuring the amplitude-frequency curve. A typical amplitude-frequency curve of QTF is shown in Figure 3.

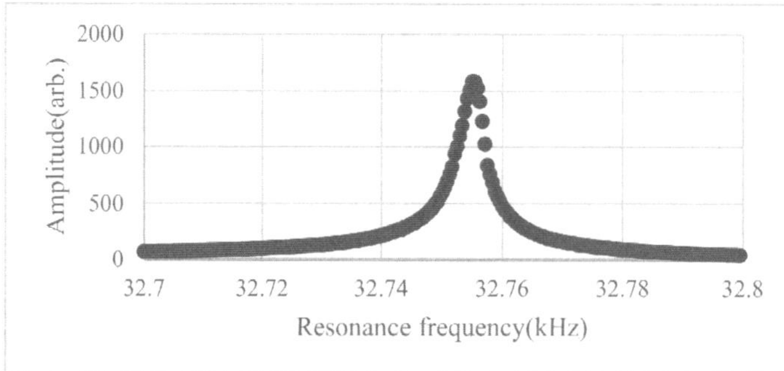

Figure 3. Typical amplitude-frequency curve of QTF.

The measurement of the amplitude-frequency curve usually requires applying excitation signals of different frequencies on the QTF to excite the mechanical vibration of the QTF prongs and then detecting the amplitude of the mechanical vibration. The experimental principle is consistent with Kim et al.'s work [14]. The measuring apparatus consists of a function generator, a preamplifier, a lock-in amplifier and a computer (in fact, a set of AFM that can use the QTF force sensor is used instead). The typical amplitude-frequency curve of the peeled QTF is shown in Figure 3. Through data analysis, the Q-factor of the peeled QTF can be calculated as follows:

$$Q = \frac{f_0}{f_2 - f_1} = 7873.$$

3. Experimental results and discussion

3.1 Influence of the position of micro ball

3.1.1 Experiment preparation

The ball used in this experiment is a lead-tin ball with a diameter of $0.4\ mm$ (tin 63% lead 37%). The micro ball is glued at different positions on the outer wall of the QTFs. When the ball is closest to the front end of the fork, the corresponding position of the ball is marked as $0\ mm$. In this experiment, 7 QTFs with the same specification were used, and the micro ball with the same diameter was bonded at 7 different positions on the outer side of the QTF prong. The photos of the QTFs after the micro ball was bonded are shown in Figure 4. From left to right and from top to bottom, the QTFs are numbered as (a)-(g). At the same time, the ball position in (a) is marked as 0 mm nearby and the ball positions in (b) to (g) are taken as a reference and marked as the actual position or planned position of these balls as shown below in Figures 5 and 6.

For the above 7 QTFs, the amplitude-frequency curve is measured before the micro ball is pasted (empty QTF) and after the micro ball is pasted. Thus, the resonance frequency and Q-factor of the QTFs can be acquired.

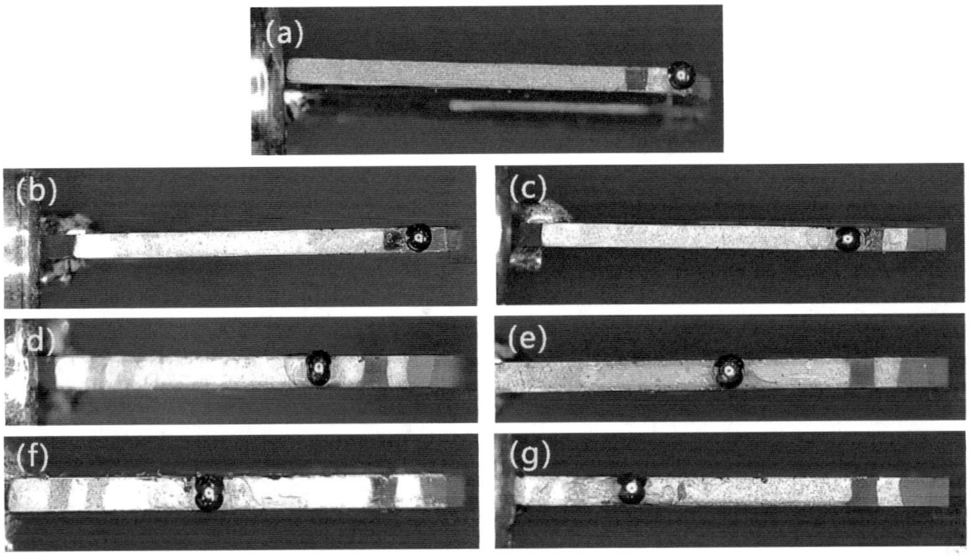

Figure 4. Photos for the micro balls to be glued on 7 different positions on the outside of the QTFs. (From left to right and from top to bottom, the QTFs are numbered as (a)-(g), while the ball position in (a) is marked as 0 mm nearby and that for (a)-(g) are taken this as a reference.)

3.1.2 Experimental result and discussion

Using the acquired data, the relationship between the resonance frequencies, including that of the empty QTFs without the micro ball and that of the QTFs with a micro ball on different positions of its prong, and their actual position are shown in Figure 5. For empty QTFs without micro balls, it can be seen from the left figure in Figure 5 that the resonance frequencies of all QTFs are between $32.7308\ kHz$ and $32.7567\ kHz$. The difference between the resonance frequencies of each QTF is small, with the maximum not exceeding $30\ Hz$.

After sticking a micro ball on the QTF prong, it can be seen from the right figure in Figure 5 that the resonance frequency of the QTF will change significantly. On the one hand, the resonance frequency of the QTF will decrease after pasting the micro ball. The closer the position of the micro ball is near to the free end of the QTF prong, the more the resonance frequency decreases; The lowest reduction in the measurement results is $29.832\ kHz$, which is about 8.9% of that of the empty ones. This experimental result is consistent with the classical model of spring vibration, i.e., Formula (1), because the increase of the effective mass of the cantilever will lead to the reduction of the resonance frequency. On the other hand, the closer it is to the fixed end of the QTF prong, the fewer changes in the resonance frequency.

The above experimental results also show that, although the actual mass of the pasted ball is basically the same, the closer position of the pasted ball to the fixed end of the QTF prong is, the smaller increase of the effective mass in Formula (4) will be, and the smaller impact on the resonance frequency will be. So, it can be inferred that whenever the position of the additional mass ball on the QTF prong is different (that is, the distance from the free end is different), its effective mass should also be different, while the classical spring vibration model or formula is used to calculate the resonance frequency.

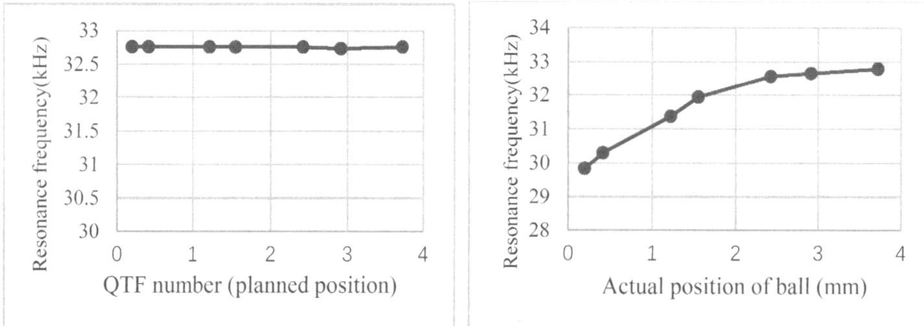

Figure 5. The resonance frequency of QTF affected by the position of the micro ball
(Left: for empty QTF without the micro ball; right: after sticking a micro ball)

Similarly, according to the acquired data, the relationship between the Q-factor, including that of the empty QTF without the micro ball and that of the QTF with a micro ball, and the position of the micro ball are shown in Figure 6. For empty QTFs without the micro ball, it can be seen from the left figure in Figure 6 that the Q-factors of all QTFs are between 6000 and 8000, and the differences between the Q-factor of each QTF are small.

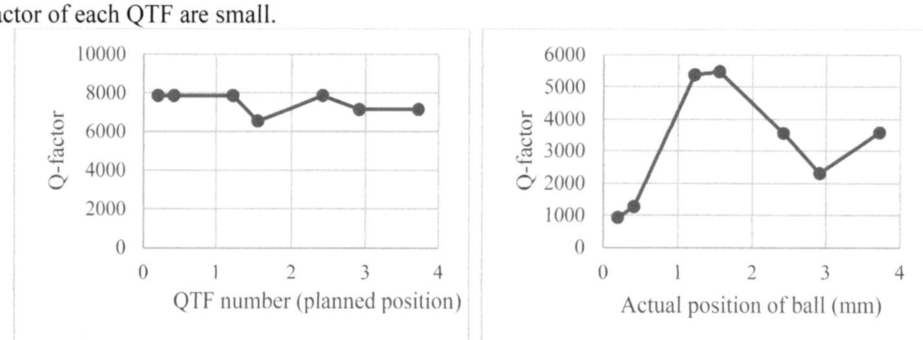

Figure 6. The Q-factor of QTF affected by the position of the micro ball.
(Left picture: for empty QTF without the micro ball; right picture: after sticking a micro ball)

After sticking a micro ball on the QTF prong of the QTF, it can be seen from the right figure in Figure 6 that the Q-factor of the QTF will change significantly. On the one hand, the Q-factor of the QTF will obviously decrease after the micro ball is pasted. In this measurement, the Q-factor of QTF is reduced to between 1000 and 5500. On the other hand, the Q-factor decreases more obviously at the position where the micro ball is pasted at the side near the free end of the QTF prong, with a minimum Q-factor of around 1000, giving a maximum reduction of about 80%. This experimental result shows that to ensure a high Q-factor when designing a QTF-based sensor, the effective mass of the additional parts on the QTF prong should be avoided or reduced as far as possible. Otherwise, one should make the attachment stick away from the free end of the QTF prong as far as possible.

3.2 Influence of the size of additional mass

3.2.1 Experiment preparation

In this experiment, micro balls with different masses (diameter) were glued to the same position on the side of the QTF prong to study their influence on the resonance sensing characteristics of the QTF. The photos of QTFs after bonding the ball are shown in Figure 7.

The balls used in this experiment is still lead-tin balls (tin 63% lead 37%), with 6 diameters: 0.6 mm, 0.5 mm, 0.4 mm, 0.3 mm, 0.25 mm, and 0.2mm. The glue used for bonding is the liquid glue that dries slowly in 100 seconds.

For the above 6 QTFs, the amplitude-frequency curve is measured before the micro ball is pasted (empty QTF) and after the micro ball with a different ball diameter is pasted. Thus, the resonance frequency and Q-factor of the QTFs can be acquired.

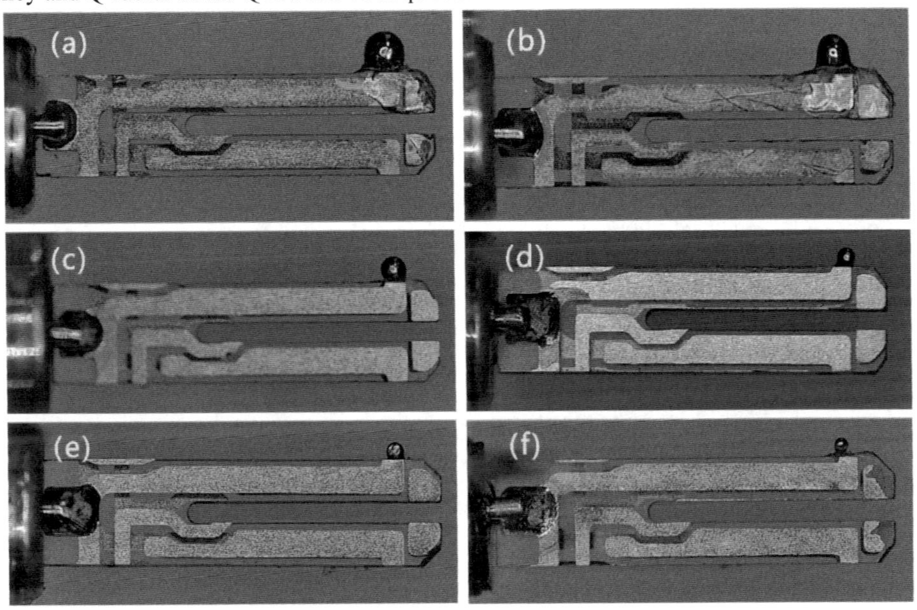

Figure 7. Photos of QTF after bonding micro balls with different diameters (different mass). QTFs are numbered as (a)-(f) from left to right and from top to bottom)

3.2.2 Experiment results and discussion

Using the acquired data, the relationship between the resonance frequencies, including that of the empty QTF without pasting a micro ball and that of the QTF with pasting a micro ball of different mass, and the mass of the micro ball are shown in Figure 8. For the empty QTF without a micro ball, it can be seen from the left figure in Figure 8 that the resonance frequency of all QTFs is between 32.7475 kHz and 32.8375 kHz. The difference between the resonance frequencies of each QTFs is small.

After sticking a micro ball on the QTF prong, it can be seen from the right figure in Figure 8 that the resonance frequency of the QTF will change significantly. The resonance frequency of the QTF will decrease after pasting a micro ball. The larger mass of the paste ball is, the more the resonance frequency decreases; The minimum reduction in the measurement results is 25.500 kHz. This experimental result is consistent with the classical model of spring vibration, i.e., Formula (5).

The relationship between the resonance frequency of the QTF and the mass (diameter) of the ball can be further analysed to give a quantitative relationship.

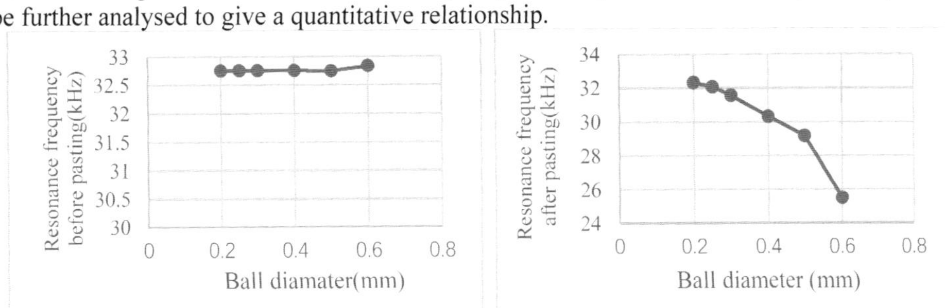

Figure 8. The relationship between the resonance frequency of QTF and ball mass (diameter). (Left: before pasting; right: after pasting)

Similarly, according to the acquired data, the relationship between the Q-factor, including that of the empty QTF without the micro ball and that of the QTF with the micro ball, and the mass (diameter) of the micro ball is shown in Figure 9. As for the empty QTF without a micro ball, it can be seen from the left figure in Figure 9 that the Q-factor of all QTFs is between 7150 and 7860, and the difference between the Q-factors of each QTF is small.

After sticking a micro ball on the QTF prong, it can be seen from the right figure in Figure 9 that the Q-factor of the QTF will change significantly. On the one hand, after pasting the micro ball, the Q-factor will obviously decrease. In this measurement result, the Q-factor of QTF is reduced to a minimum value of about 2.6. The heavier the ball attached to the QTF is, the greater reduction of the Q-factor decreases. This experimental result shows that to ensure a high Q-factor when designing the sensor based on QTF, the mass of the attached object should be minimised as far as possible.

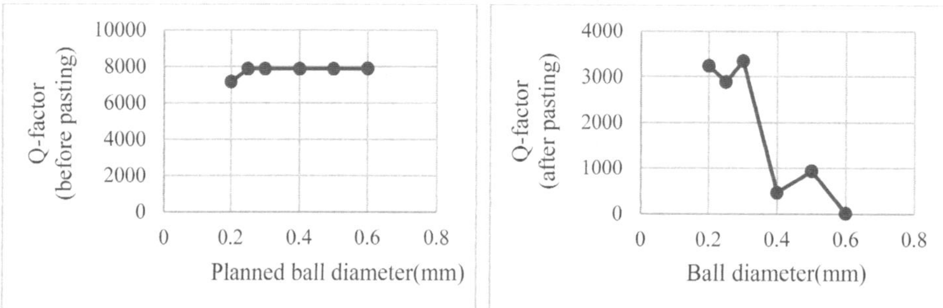

Figure 9. Relationship between Q-factor of QTF and ball mass (diameter).
(Left: before pasting; right: after pasting)

4. Conclusion

Additional mass on the prong has a critical influence on the mechanical resonance and sensing characteristics of the QTF sensor. Employing the Q-factor as the key factor of the sensor sensitivity, the mechanical resonance and sensing characteristics of QTF with an additional mass on its prongs are studied by theoretical analysis and experiments.

First, the resonance frequency, Q-factor and their influencing factors are analysed and calculated using the classical mechanical model. The additional mass of the tip will affect the resonance frequency and Q-factor of QTF, with the quantitative relationship being analysed and inferred.

Secondly, our experimental results show that the additional mass on QTF has a significant impact on the resonance frequency of its mechanical vibration. The greater the added mass is, the lower the resonance frequency is. The closer the position of the additional mass to the free end of the QTF prongs is, the greater the influence on the resonance frequency will be. There is a roughly linear relationship between the relative variation of the resonant frequency and the size of the additional mass.

Thirdly, our experimental results show that the additional mass on QTF has a significant impact on its Q-factor. The greater the effective mass is, the more the Q-factor will decrease. Reducing the additional mass on the QTF prong can significantly improve its Q-factor, thus improving the sensing performance of the QTF force sensor.

The research results in this paper are conducive to a deeper understanding of the QTF sensor. They can also provide a quantitative reference for the design and optimisation of the QTF force sensor, especially for its application in AFM.

Acknowledgements
The authors gratefully acknowledge the financial support of the National Natural Science Foundation of China (Grant No. 61871410).

References

[1] J.-M. Friedt and É. Carry. 2007 Introduction to the quartz tuning fork. *Am. J. Phys.* **75** 415
[2] XU Jun and YANG Fan and SUN Ming-xiao. 2020 Design of Quartz Tuning Fork Humidity Sensor Based on Polyvinyl Alcohol Film. *Instrument Technique and Sensor* (in Chinese). **11** 5
[3] XU Jun, MENG Yue-xia, LIU Chun-hua and MA Jing. 2017 Test System for Quartz Tuning Fork Temperature Sensor. *Journal of Harbin University of Science and technology* (in Chinese). **22** 44
[4] ZHANG Mi, CHEN De-hua, LIU Shu-jian and Li Ping. 2022 Study on Fluid Density and Viscosity Measured by Quartz Tuning Fork Sensor. *Instrument Technique and Sensor* (in Chinese). **21** 19
[5] Dong Lei, Li Shangzhi, Shang Zhijin and Wu hongpeng. 2022 Advances in Quartz tuning fork based photoacoustic gas sensing. *Journal of Henan Normal University (Ntural Science Edition)* (in Chinese) **50** 16
[6] Jean-Ochin Abrahamians, Laurent Pham Van and Stéphane Régnier. 2016 Contributed Review: Quartz force sensing probes for micro-applications. *Rev. Sci. Instrum.* **87** 071502
[7] Yingxu Zhang, Yingzi Li, Zihang Song, Rui Lin, Yifu Chen and Jianqiang Qian. 2018 A High-Q AFM Sensor Using a Balanced Trolling Quartz Tuning Fork in the Liquid. *Sensors*. **18** 1628
[8] Ke Xu, Chenghao Ji1, Fanan Wei and Zhushan Zhai. 2021 Recent Progress in Self-sensing Probe Technology in Atomic Force Microscope. *Current Nanoscience*, **17** 808-18
[9] Zhichao Wu, Tong Guo, Ran Tao, Leihua Liu, Jinping Chen, Xing Fu and Xiaotang Hu. 2015 A Unique Self-Sensing, Self-Actuating AFM Probe at Higher Eigenmodes. *Sensors*. **15** 28764
[10] A. Castellanos-Gomez1, N. Agraït and G. Rubio-Bollinger. 2009 Dynamics of quartz tuning fork force sensors used in scanning probe microscopy. *Nanotechnology*. **20** 215502
[11] Yuan-Liu Chen, Yanhao Xu, Yuki Shimizu, Hiraku Matsukuma and Wei Gao. 2018 High quality-factor quartz tuning fork glass probe used in tapping mode atomic force microscopy for surface profile measurement. *Meas. Sci. Technol.* **29** 065014
[12] Hua Bao-Cheng, Qian Jian-Qiang, Wang Xi and Yao Jun-En. 2011 Mechanical model of tuning forks used in scanning probe microscopes. *Acta Phys. Sin.* (In Chinese) **60** 040702
[13] Yuya Yamada, Takashi Ichii, Toru Utsunomiya and Hiroyuki Sugimura. 2020 Visualizing polymeric liquid/solid interfaces by atomic force microscopy utilizing quartz tuning fork sensors. *Japanese Journal of Applied Physics*. **59** SN1009
[14] Jongwoo Kim, Donghyun Won, Baekman Sung, Sangmin An and Wonho Jhe. 2014 Effective stiffness of qPlus sensor and quartz tuning fork. *Ultramicroscopy*. **141** 56
[15] Xiaofei Zhang, Fengli Gao and Xide Li. 2018 Sensing Performance Analysis on Quartz Tuning Fork-Probe at the High Order Vibration Mode for Multi-Frequency Scanning Probe Microscopy. *Sensors*. **18** 336
[16] Boon Ping Ng, Ying Zhang, Shaw Wei Kok and Yeng Chai Soh. 2009 Improve performance of scanning probe microscopy by balancing tuning fork prongs. *Ultramicroscopy*. **109** 291–5
[17] Danish Hussam, Jianmin Song, Hao Zhang, Xianghe Meng, Wen Yonbing and Hui Xie. 2017 Optimizing the quality factor of quartz tuning fork force sensor for atomic force microscopy: impact of addition mass and mass rebalance. *IEEE Sensors Journal*. **17** 2797

Journal of Physics: Conference Series **2561** (2023) 012005 doi:10.1088/1742-6596/2561/1/012005

Study on Calibration of Oxygen Transmission Rate Measurement Apparatus for Plastic Film and Sheeting

Yang Xu[1], Ying Liu[1], Tingting Ren[1*], Lingjian Zeng[1], and Xianhong Li[1]

[1] Chongqing Academy of Metrology and Quality Inspection, Chongqing, 401123, China

[*]Corresponding author's e-mail: cd@cqjz.com.cn

Abstract. The oxygen transmission rate measurement apparatus for plastic film and sheeting is mainly used to measure the oxygen transmission rate and oxygen transmission coefficient of plastic film and sheeting. Oxygen transmission tests can analyze and solve the problems caused by oxygen, such as oxidation deterioration. At present, there is no relevant calibration specification or method for oxygen transmission rate measurement apparatus for plastic film and sheeting in China. This article presents a novel calibration method to evaluate the measurement accuracy and temperature control performance of the apparatus by determining corresponding environmental conditions, calibration standards, and calculation equations. The calibration results show that the calibration method is scientific and practical for the metrological performance evaluation of oxygen transmission rate measurement apparatus for plastic film and sheeting.

1. Introduction

The oxygen transmission rate measurement apparatus for plastic film and sheeting is mainly used to measure the oxygen transmission rate and oxygen transmission coefficient of plastic film and sheeting, and other packaging materials and plastic products. It is widely used in packaging and material manufacturing. According to the working methods, there are two types of oxygen transmission rate measurement apparatus: one uses a pressure sensor, and the other a coulometric sensor to measure the amount of oxygen that permeates through a test specimen.[1]

Figure 1 shows an example of an apparatus for using a pressure sensor. A test specimen is mounted in a gas transmission cell to form a sealed barrier between two chambers. The lower-pressure chamber is evacuated, followed by the evacuation of the higher-pressure chamber. A gas is introduced into the evacuated higher-pressure chamber and permeates into the lower-pressure chamber. The amount of gas that permeates through the specimen is determined by the increase in pressure on the lower-pressure side. The apparatus consists of a gas transmission cell designed to allow the oxygen to permeate through a specimen, a pressure sensor to detect the pressure change due to the permeation of the oxygen through the specimen, an oxygen feeder to supply the oxygen to the transmission cell, a cell volume-control device to extend the transmission rate measurement range and a vacuum pump capable of producing a vacuum better than 10 Pa.[2]

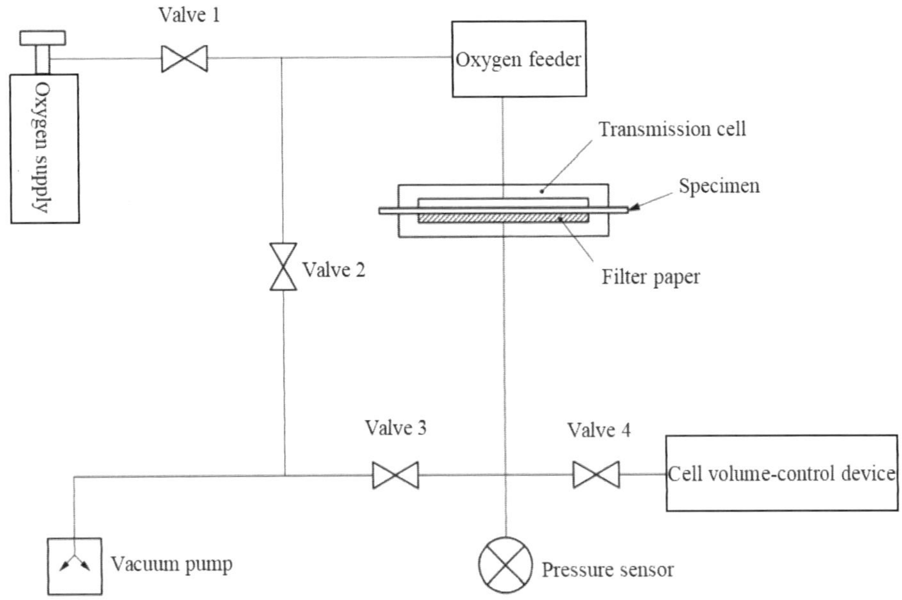

Figure 1. Example of oxygen transmission rate measurement apparatus using a pressure sensor

Figure 2 shows a typical apparatus using a coulometric sensor. A test specimen is mounted in a gas-transmission cell to form a sealed barrier between two chambers. One chamber is slowly swept with a carrier gas. The second chamber is fed with the test gas. The total pressure is identical (atmospheric) in each chamber, but since the partial pressure of the test gas is higher in the second chamber, the test gas permeates through the barrier into the carrier gas in the first chamber. The test gas which permeates through the specimen is carried by the carrier gas to a sensor, the nature of which depends on the material under test and the test gas used. The valve (12) bypasses the sensor, thus avoiding the cell being swamped with air when the test specimen is being placed in the cell. The valves (7) immediately upstream of the transmission cell allow either carrier gas or test gas to be passed into chamber A of the cell. The catalyst bed is intended to eliminate any traces of oxygen which might be present in the carrier gas.[3-4]

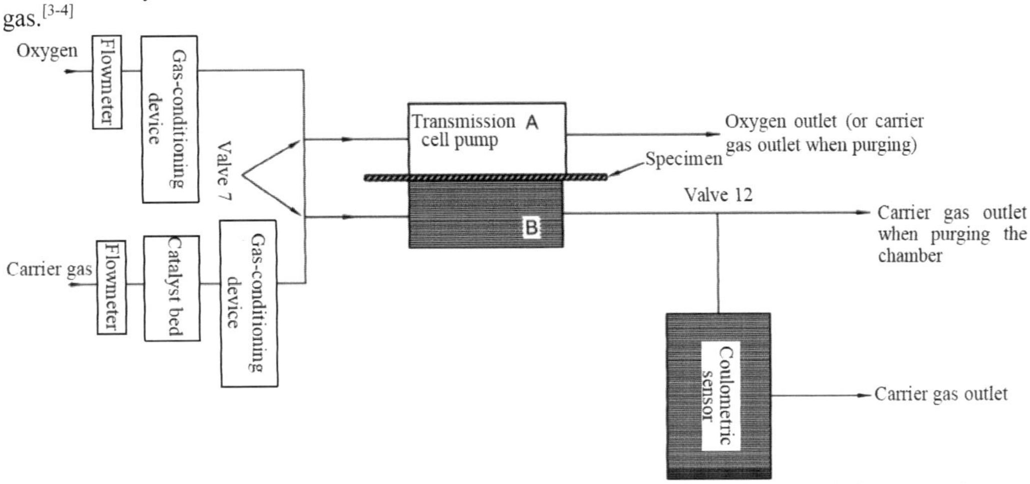

Figure 2. Typical example of apparatus for the determination of oxygen-transmission rate using a coulometric sensor

The oxygen transmission test belongs to the barrier performance test, and the barrier performance of plastic film and sheeting can greatly affect the product life. Oxygen transmission tests can analyze and solve the problems caused by oxygen, such as oxidation deterioration[5-11]. However, there is no calibration specification for oxygen transmission rate measurement apparatus for plastic film and sheeting in China at present, and the metrological performance of the apparatus cannot be ensured. Therefore, it is urgent to develop an effective calibration method that can evaluate the metrological performance of oxygen transmission rate measurement apparatus for plastic film and sheeting, which can ensure the reliability of the apparatus in daily use.

2. Calibration conditions

2.1. Environmental conditions
During the whole measurement process, the temperature should be (21~25) °C, and the humidity should be no more than 80%RH.

2.2. Metrological standards
Standard films of oxygen transmission rate are needed to calibrate the oxygen transmission testing ability of the measurement apparatus: the standard film with a transmission rate of no more than $20cm^3/(m^2 \cdot 24\ h \cdot 0.1MPa)$, the standard film with a transmission rate of $(20\sim100)cm^3/(m^2 \cdot 24\ h \cdot 0.1MPa)$, and the standard film with a transmission rate of more than $100cm^3/(m^2 \cdot 24\ h \cdot 0.1MPa)$. The relative expanded uncertainty of the standard films should be no more than 6% (k=2). A standard thermometer with a maximum permissive error of no more than ±0.3°C is needed to calibrate the temperature parameter of the measurement apparatus.

3. Calibration method

3.1. Relative indication error of oxygen transmission rate
A standard film of an oxygen transmission rate of no more than $20cm^3/(m^2 \cdot 24\ h \cdot 0.1MPa)$ is selected and packaged in a selected test chamber of the calibrated measurement apparatus according to the operating instruction of the calibrated apparatus and the standard film. The temperature of the apparatus is set at 23 °C, the humidity at 0%, and the measurement status to "pkg" packaging mode. The oxygen transmission rate of the standard film is measured three times, followed by the measurement of the standard films of the transmission rate of $(20\sim100)cm^3/(m^2 \cdot 24\ h \cdot 0.1MPa)$ and more than $100cm^3/(m^2 \cdot 24\ h \cdot 0.1MPa)$ in the selected chamber.

The rest chambers of the apparatus are calibrated according to the method above. The relative indication error of the oxygen transmission rate δ_{OTR} is calibrated with each standard film in different chambers according to Equation (1).

$$\delta_{OTR} = \frac{\overline{OTR} - OTR_s}{OTR_s} \times 100\% \tag{1}$$

Where $\overline{OTR}$ is the mean of oxygen transmission rate indications of the calibrated apparatus, and OTR_s is the standard value of the standard film.

The maximum value of all the calculated results is taken as the relative error of the oxygen transmission rate measurement apparatus.

3.2. *Measurement repeatability of oxygen transmission rate*
The repeatability of oxygen transmission of the apparatus are calibrated according to the method presented in Chapter 3.1, and is calculated according to Equation (2).

$$s_{OTR} = \frac{OTR_{\max} - OTR_{\min}}{1.69 \times \overline{OTR}} \times 100\% \quad (2)$$

Where OTR_{max} is the maximum of all three indication values of the calibrated apparatus, while OTR_{min} is the minimum.

The maximum value of all the calculated results is taken as the measurement repeatability of the oxygen transmission rate measurement apparatus.

3.3. Test chamber temperature indication error

The temperature of the test chamber of the apparatus is set at 23°C. The probe of the standard thermometer is fixed in the temperature detection port of the test chamber. After the temperature of the test chamber is stable, the temperature of the test chamber is measured every 5 minutes, and a total of six temperature measurements are obtained.

The rest chambers of the apparatus are calibrated according to the method above. The test chamber temperature indication error Δt is calculated according to Equation (3).

$$\Delta t = 23 - \bar{t} \quad (3)$$

Where $\bar{t}$ is the mean of six temperature measurements.

Take the maximum value of all the calculated results as the test chamber temperature indication error of the oxygen transmission rate measurement apparatus.

3.4. Test chamber temperature stability.

The temperature of the test chamber of the apparatus is set at 23°C. The probe of the standard thermometer is fixed in the temperature detection port of the test chamber. After the temperature of the test chamber is stable, the temperature of the test chamber is measured every five minutes, and a total of six temperature measurements are obtained.

The rest chambers of the apparatus is calibrated according to the method above. The test chamber temperature stability δ_t is calculated according to Equation (4).

$$\delta t = t_{\max} - t_{\min} \quad (4)$$

Where t_{max} is the maximum of all six temperature measurements, while t_{min} is the minimum.

Take the maximum value of all the calculated results as the test chamber temperature stability of the oxygen transmission rate measurement apparatus.

4. Experimental verification of calibration method

Five typical types of oxygen transmission rate measurement apparatus for plastic film and sheeting are set as the calibration subjects for the verification experiment. The calibration results are demonstrated in Table 1.

Table 1. Calibration results of five typical types of oxygen transmission rate measurement apparatus for plastic film and sheeting.

Type	Parameters		Result
Y310	Oxygen transmission rate	Relative indication error (%)	-16.7
		Repeatability (%)	2.9
	Test chamber temperature	Indication error (°C)	0.8

		Stability (°C)	0.4
8101e	Oxygen transmission rate	Relative indication error (%)	15.5
		Repeatability (%)	2.8
	Test chamber temperature	Indication error (°C)	0.9
		Stability (°C)	0.3
DRK310 OTR	Oxygen transmission rate	Relative indication error (%)	13.7
		Repeatability (%)	4.5
	Test chamber temperature	Indication error (°C)	-0.6
		Stability (°C)	0.2
GPT-201	Oxygen transmission rate	Relative indication error (%)	-15.2
		Repeatability (%)	2.3
	Test chamber temperature	Indication error (°C)	0.7
		Stability (°C)	0.3
GTR-V3	Oxygen transmission rate	Relative indication error (%)	10.5
		Repeatability (%)	1.7
	Test chamber temperature	Indication error (°C)	-0.8
		Stability (°C)	0.4

As we can see from Table 1, the maximum relative indication error of the oxygen transmission rate is no more than ±20%, the maximum repeatability of the oxygen transmission rate is no more than 5%, the maximum indication error of the test chamber temperature is no more than ±1.0°C, and the maximum stability of test chamber temperature is no more than 0.5°C, which can meet the requirement of the enterprise standards of manufacturers. The feasibility and applicability of the calibration method have been proven.

5. Conclusion

This article presents a novel calibration method for oxygen transmission rate measurement apparatus for plastic film and sheeting to evaluate the measurement accuracy and temperature control performance of the apparatus by means of determining corresponding environmental conditions, calibration standards, and calculation equations. The calibration results show that the calibration method is scientific and practical for the metrological performance evaluation of oxygen transmission rate measurement apparatus for plastic film and sheeting. The method can greatly improve the inspection accuracy of the material transmittance and help users efficiently find the specific plastic film and sheeting according to their special needs. Due to the improvement of the accuracy of oxygen transmission rate measurement apparatus, quality inspection institutions can monitor the safety and stability of product packaging and resolve disputes between consumers and producers caused by packaging quality problems more efficiently.

References

[1] Cao, L, et al. (2022) Research on the Comparison and Analysis of Oxygen Transmission Rate Tester and Related Data [J]. Modern Information Technology, (18)04:47-50
[2] Xu, B, et al. (2022) GB/T 1038.1 Plastics--Film and sheeting--Determination of gas-transmission rate--Part 1:Differential-pressure methods [M]. Standards Press of China, Beijing.
[3] Xu, B, et al. (2022) GB/T 1038.2 Plastics--Film and sheeting--Determination of gas-transmission rate--Part 2:Equal-pressure methods [M]. Standards Press of China, Beijing.

[4] ASTM. Designation. (2007) ASTM F 1927-07 Standard test method for determination of oxygen gas transmission rate, permeability and permeance at controlled relative humidity through barrier materials using a coulometric detector [S]. ASTM, Philadelphia, PA.
[5] Xing, W, et al. (2015) Effects of different temperatures and humidity on oxygen transmission rate of different structures of the milk powder packaging film [J]. Food Science and Technology, (02)40:47-50
[6] Lv, Y, N, et al. (2012) Research on the temperature and humidity control technology of the isobaric oxygen permeability tester [J]. China Packaging, 4:59-63.
[7] Wu, Q, T. (2008) Flexible Packaging Structure Design and Process Design [M]. Printing Industry Press, Beijing.
[8] Gao, Q. (2014) Study of the water vapor transmission rate for the organic/inorganic/organic hybrid barrier film [J]. Journal of Optoelectronics·Laser, (25)09: 1721-1726.
[9] Peng, R. (2020) Research on the measurement method of water vapor transmittance of OLED encapsulation film [J]. Electronics & Packaging, (20)02: 26-30.
[10] Li, C, et al. (2017) Study on test system for water vapor transmission rate in composite thin film encapsulation [J]. Journal of Technology, (17)01: 27-31.
[11] Mao, S, et al. (2019) Study on calibration method of the oxygen gas permeability characteristics. Journal of Test and Measurement Technology, (06)33: 480-483.

Design of structure and control system based on achromatic rotating biprism

Yue Shi [1a], Jiahong Zheng [2, b*], Lei Ming [3] and Jiahou Dong [4]

[1] College of Mechatronics Engineering, Shaanxi University of Science and Technology, Xi'an City, Shaanxi Province, 710016, China

[2]College of Mechatronics Engineering, Shaanxi University of Science and Technology, Xi'an City, Shaanxi Province, 710016, China

[3] College of Mechatronics Engineering, Shaanxi University of Science and Technology, Xi'an City, Shaanxi Province, 710016, China

[4] College of Mechatronics Engineering, Shaanxi University of Science and Technology, Xi'an City, Shaanxi Province, 710016, China

[a] 1029471128@qq.com, [b*]200512091@sust.edu.cn

Abstract. The achromatic rotating dual-prism beam pointing system is mostly used in space laser communication, which needs to meet the characteristics of small size and light mass. To address the problems of the large size, heavy mass and low accuracy of the conventional beam, this paper uses SOLIDWORKS to build a dual-optical wedge prism model and the structural model of the optical machine and simulates and analyses its environmental performance. The design of the control system effectively improves the beam-pointing accuracy of the whole machine. A reference value is provided for the study of rotating biprism.

1. Introduction

The research of beam pointing control technology is of considerable significance for space tracking, targeting, space communication, etc. [1-3]. The traditional beam pointing mechanism includes a universal rotating frame structure and fast reflector structure, which has the disadvantages of low pointing accuracy, large size, large inertia, slow response frequency and poor dynamic performance [4-6,7]. Therefore, in the modern beam-pointing control mechanism research, the two structures will be combined; the development of the rotating dual-prism beam-pointing system can effectively compensate for the shortcomings of the traditional beam-pointing mechanism [8,9]. Thus, the characteristics of a large scanning range, high pointing accuracy, simple structure and high response frequency are realized. The working principle of the rotating dual prism system is that the two prisms are rotated independently by co-axis to achieve beam or view axis pointing adjustment; at present, the rotating dual prism has shown wide application prospects in beam scanning, long-range target aiming and tracking, etc. [10,11].

2. Optical machine structure design analysis

2.1. Optical machine structure design

According to the design requirements of the project regarding the dimensional part of the structure, and considering that the structure will not block the light, the appearance design results are shown in Figure

1.

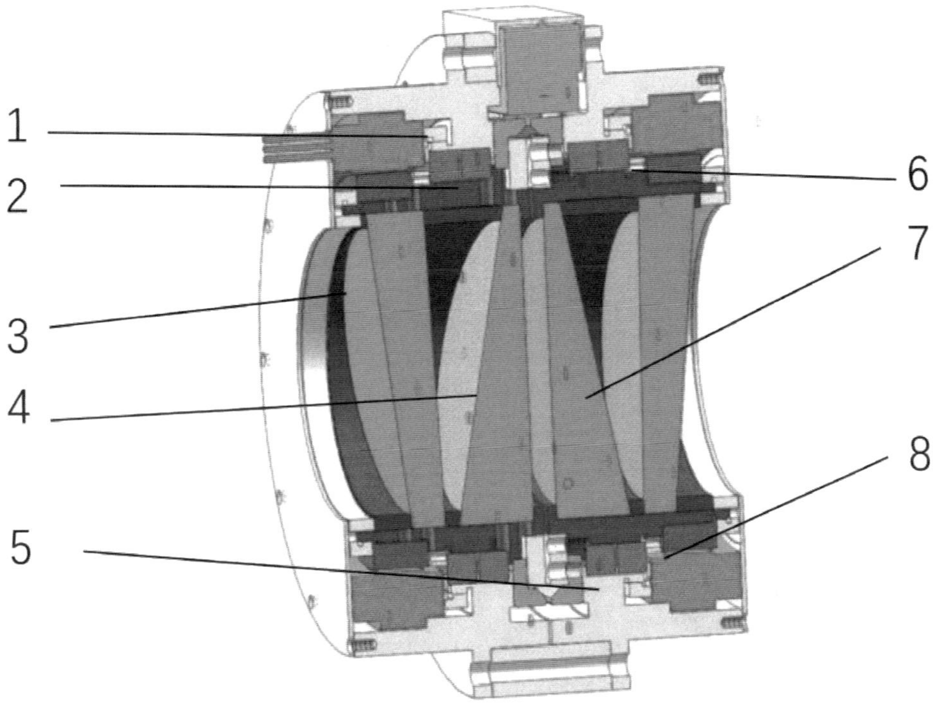

1. Motor stator 2. Motor rotor 3. Housing sealing plate 4. Achromatic prism 5. Bearing 6. Round grating 7. Main prism 8. Mirror set spacer

Figure 1. Risley prism structure design profile

2.2. Environmental suitability analysis

To ensure the stability of the achromatic rotating biprism system in the process of use, finite element analysis is required for its environmental adaptation. The materials used in the achromatic rotating biprism lens are monocrystalline silicon and monocrystalline germanium. The materials used in the achromatic rotating biprism optical structure are titanium alloy and bearing steel. Considering the thermal matching between the structural material and the optical material, and considering the rigidity of the structural material, processability and successful application cases, the titanium alloy is chosen as the material of the structural parts. The assembly gap is fixed by silicone rubber, and the properties of each material are shown in Table 1:

Table 1. Properties of monocrystalline silicon materials

Material Name	Density (kg/m^3)	Modulus of elasticity (GPa)	Poisson's ratio	Thermal conductivity ($\times 10^{-6}$/K)	Specific heat capacity W/(m·K)	Specific heat capacity (cal/(g·K))	Fracture strength (MPa)
Monocrystalline Silicon	2320	131	0.28	2.4	150	0.169	≥70
Monocrystalline Germanium	5327	103	0.28	5.7	58.61	0.08	≥90
Titanium alloy	5620	96	0.36	9.4	21.9	0.13	930
Bearing steel	7850	200	0.3	12	60.5	0.11	250
XM-31 rubber	1200	0.0345	0.38	65	5	201.5	0.83

Journal of Physics: Conference Series **2561** (2023) 012006 doi:10.1088/1742-6596/2561/1/012006

2.3. Environmental adaptation simulation

The finite element analysis simulation process is shown in Figure 2.

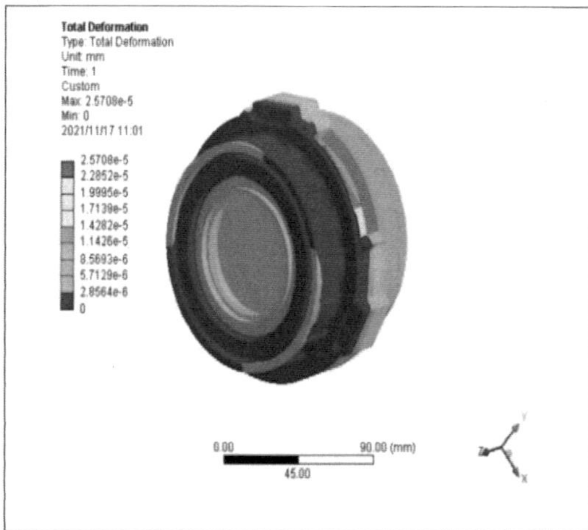

a) Single gravity impact

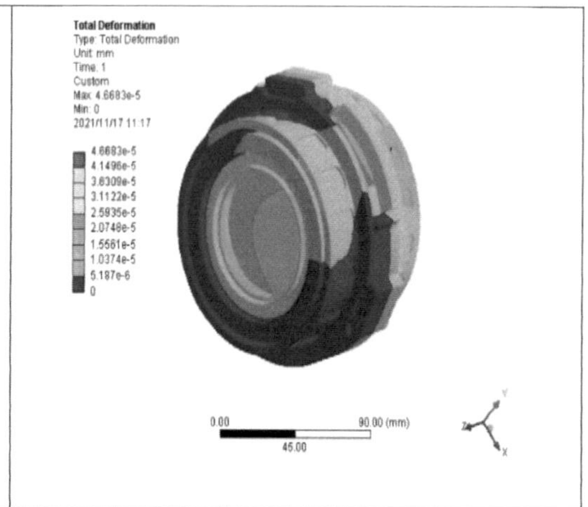

b) Forward 1.5 g impact

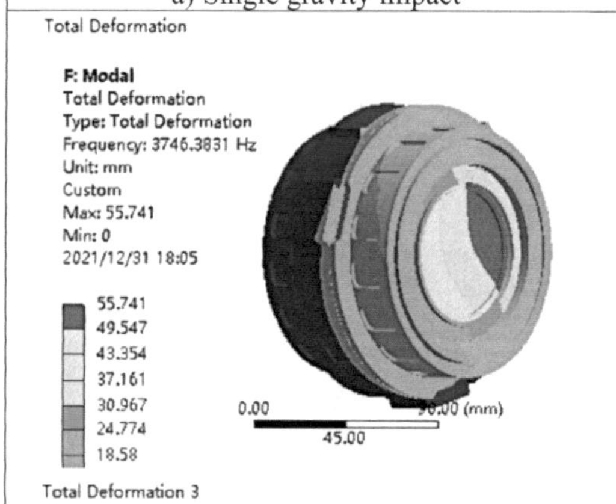

c) 3746 HZ Modal Analysis

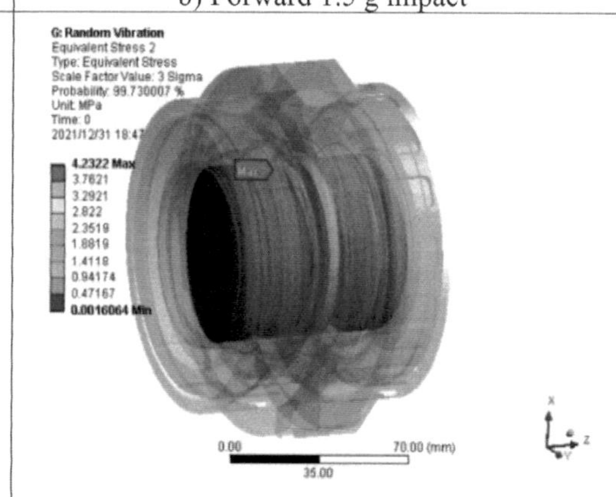

d) Mirror stress distribution

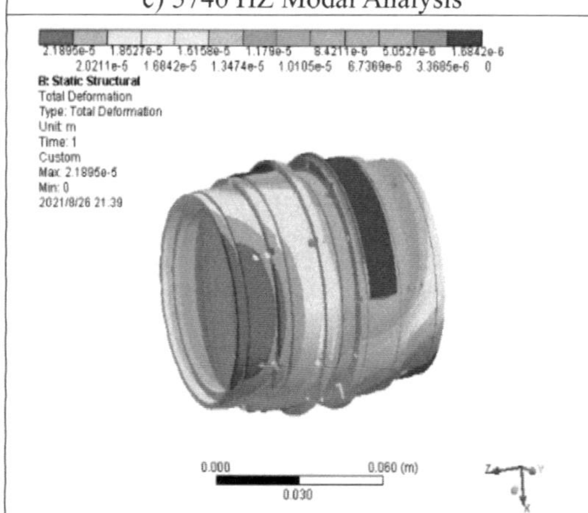

e) Extreme high temperature +70 °C

f) Extreme low temperature -50 °C

Figure 2. Finite element simulation analysis results

According to the figures, under the influence of gravity applied to the double light wedge alone, the maximum deformation of the double light wedge appears to be 2.57×10^{-5} mm. Under the influence of forward 1.5 g overload, the maximum deformation of the double light wedge appears to be 4.67×10^{-5} mm, and the stress is far less than the design requirements. The results of the modal analysis show that when the first-order frequency of the equipment reaches 3746 HZ, it is not easy to resonate with the external. The results of the modal analysis show that when the first-order frequency of the equipment reaches 3746 HZ, it is not easy to resonate with the external frequency when in use, with very high structural rigidity and dynamic response. The maximum deformation is 2.2×10^{-2} mm in the extremely high temperature +70 °C environment and 3.06×10^{-2} mm in the extremely low temperature -50 °C environment, which has less influence on the use of the structure.

3. Control design scheme

3.1. System block diagram

The controller and driver are designed as one and integrated into a PCB board with a size of 150*100. The dual prism controller adopts a unified interface for the power supply. The master controller controls the dual prism controller through the master RS422 port and monitors the status of the dual prism controller. The dual prism controller receives the camera's off-target volume data through the master RS422 interface, solves the dual prism motor angle position in real-time and outputs the command to the driver, which drives the motor to run to the target position. The dual prism controller is connected to the azimuth motor and the pitch motor through a 10 m cable, and all the connectors are wired out perpendicular to the upper surface of the controller. The system schematic is shown in Figure 3:

3.2. Dual prism controller size layout

The drive model number is confirmed as G-BEE30P100SE-E0, and the drive size parameters are shown in Figure 4. The controller is designed to control 2 units of this drive at the same time. The controller dimensions are shown in Figure 4, with an overall size of 100*150 mm and 4 M3 screws in the mounting holes. The total weight of the biprism controller is <1.5 Kg. The biprism controller dimensional drawing is shown in Figure 4. The layout of the biprismatic controller dimensions is shown in Figure 4.

3.3. G-BEE30P100SE-E0 drive size design

The dual prism controller has 7 external interfaces. All interfaces are arranged on the long side of the upper surface of the controller. The connectors are bent and plugged into the printed board type, and the direction of the output line is horizontal.

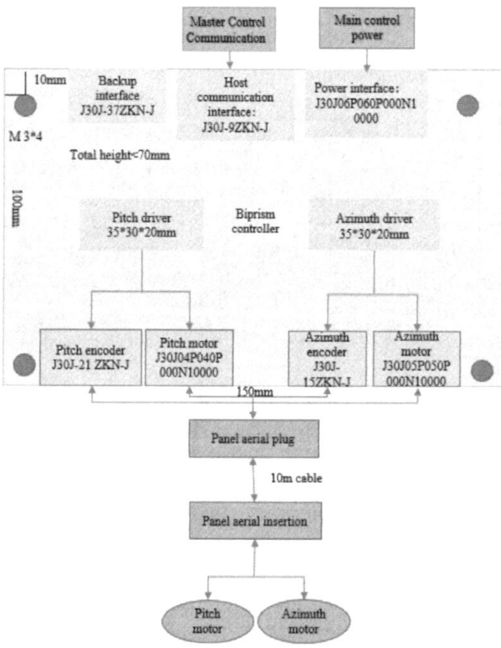

Figure 3. Block diagram of biprism control system

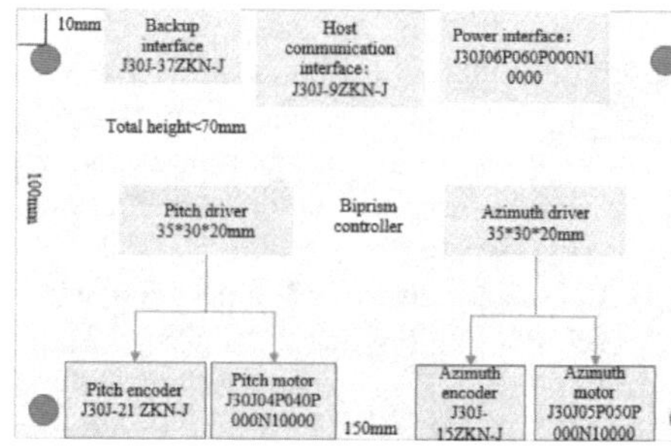

Figure 4. Controller size diagram

Figure 5. Driver dimensional drawing

The lower surface of the dual prism controller is welded with two drivers, the actual size of which is shown in Figure 5. The lower surface of the drivers is in contact with the surface of the controller mounting base plate, which is coated with heat-dissipation silicone grease to improve the heat-dissipation efficiency of the drivers.

4. Achromatic rotating dual prism beam pointing principle verification system construction

4.1. Principle verification platform introduction

The test equipment mainly includes the achromatic rotating dual prism system, target plate, laser, display, double light wedge support, high-precision lifting displacement platform and other components. To verify the beam pointing of the achromatic rotating dual prism designed in this paper, the laser beam needs to be emitted so that the beam passes through the double light wedge. The LABVIEW software is used to control the motor rotation to make the dual prism rotate and finally realize the beam. The control system software can be connected to a computer to input data directly. The completed experimental platform is shown in Figure 6.

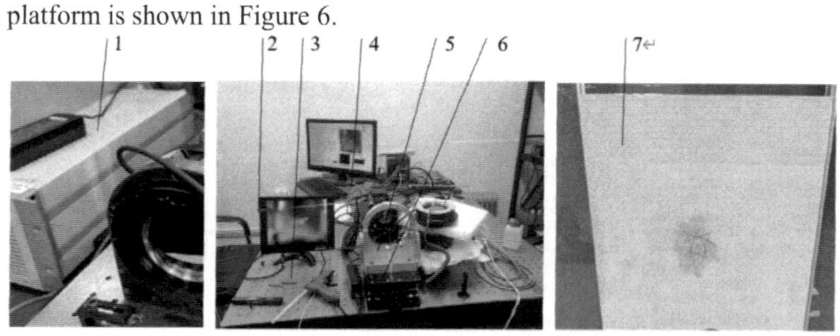

1. Laser 2. Display 3. Hot melt adhesive 4. Detector 5. Double light wedge holder 6. Panning table 7. Target plate

Figure 6. Physical picture of the experimental platform

4.2. Principle verification system platform construction process

According to the theoretical analysis described earlier, the study will verify the system beam-pointing situation. It will build the beam pointing principal verification system platform for experiments. The completed experimental platform is in Figure 6. To ensure that the system has good pointing accuracy, the beam pointing error due to external factors is reduced. The experimental platform-building process and use methods are as follows:

(1) The grid paper with a square of 1 cm^2 is used as the target plate. (2) The laser and the reflector are placed on the optical platform according to a certain position to ensure that the laser beam can enter the double prism after it is emitted. The position is corrected by using the line projector. (3) The position of the bracket needs to determine to ensure that the bracket of the centre of the circular hole and the detector lens centre in the same level. Two hair perpendicular to the edge of the bracket is used. The intersection of the hair is the bracket of the centre of the circular hole. The projector is then opened so that the light is at the intersection of the hair and through the detector centre point. After ensuring the light is in the centre of the hole, the bracket is locked with screws. The displacement platform where the detector is placed at the back end of the wedge is fixed with hot melt adhesive to prevent the position from shifting. (4) The ring pointing calibration is opened. The zero position is calibrated, and the experimental verification is prepared. Whether the zero position is calibrated is based on two points, the rotation of the two motors is firstly controlled to achieve beam deflection and observe the spot direction so that the spot position coincides with the bottom of the target plate. Then, the relative position of the prism lens of the double light wedge and the optical machine structure is observed. The highest point is at the highest point of the outer circle of the optical machine structure.

After setting up the platform of the principle verification system, the system works. This set of designs works as follows: the target is at an unknown point in space → the system scans and finds it → the reading head reads the point coordinates → the coordinates are fed back to the control system → the control system adjusts the motor rotation angle (θ_1, θ_2) → the final beam is pointed to this point.

4.3. Experimental data measurement

There are countless points on the target surface. Since measuring all the points on the target surface one by one during the experimental operation is impossible, 20 points on the X-axis are selected for measurement on average in this experiment. The rest of the measurement method is similar. The axis of the target surface is set in the following way. Facing the target surface, the centre of the target surface is the origin of the coordinate system, which is recorded as point 0. The horizontal direction of the point is the X-axis, and the vertical direction is the Y-axis. The Z-axis is the distance from the position of the prism beam to the target plate, which is a fixed value recorded as d, d = 1064 mm. The X-axis pointing accuracy in the full field of view is measured, shown in Table 2.

Table 2. X-axis pointing accuracy in the full field of view

No.	Theoretical pointing angle (mrad)	Measured pointing angle (mrad)	No.	Theoretical pointing angle (mrad)	Measured pointing angle (mrad)
1	-500,0	-500.2,0.1	11	0,0	0.2,0.2
2	-450,0	-450.1,0.1	12	50,0	50.3,0.2
3	-400,0	-400.1,0.1	13	100,0	100.1,0.1
4	-350,0	-350.2,0.0	14	150.0	150.2, -0.1
5	-300,0	-299.8,0.1	15	200,0	200.1,0.0
6	-250,0	-249.9,0.1	16	250,0	250.2,0.1
7	-200,0	-200.1,0.1	17	300,0	300.1,0.1
8	-150,0	-150.1,0.2	18	350,0	350.3,0.3
9	-100,0	-100.1,0.1	19	400,0	400.2,0.4
10	-50,0	-50.1,0.1	20	450,0	450.1,0.3

By calculating the measured data in the table, the average pointing accuracy is 0.19mrad; it meets the design index requirements.

5. Conclusion

In this paper, through the design of the dual-light wedge prism model and the structural model of the optical machine, the stability and adaptability of the structure under a complex environment are analysed by finite element simulation. It is concluded that the structure meets the design index and application requirements. It meets the design requirements of the operating temperature: -50~+70 °C and meets the requirements of shock and vibration use under the GJB150.16A-2009 environment. Secondly, the control system of the whole machine is designed. The average value of X-axis full-field beam-pointing accuracy is 0.19 mrad verified by experiment. The design of this control system effectively improves the beam-pointing accuracy of the whole machine.

References

[1] Lu S, Gao M, Yang Y, et al. Inter-satellite laser communication system based on double Risley prisms beam steering[J]. Applied optics, 2019, 58(27): 7517-7522. https://doi.org/doi:10.1364/ao.58.007517.

[2] Lavigne V, Ricard B. Fast Risley prisms camera steering system: calibration and image distortions correction through the use of a three-dimensional refraction model[J]. Optical Engineering, 2007, 46(4): 043201-043201-10. https://doi.org/doi:10.1117/1.2722754.

[3] Davis Jr R A. A new look at barrier-inlet morphodynamics[J]. Journal of Coastal Research, 2013 (69): 1-12. https://doi.org/doi:10.2112/SI_69_2.

[4] Dillon T E, Schuetz C A, Martin R D, et al. Nonmechanical beam steering using optical phased arrays[C]. Unmanned/Unattended Sensors and Sensor Networks VIII. SPIE, 2011, 8184: 98-108. https://doi.org/doi:10.1117/12.898356.

[5] Wang J, Ge Y, Chen Z D, et al. Analytic solution for double optical metasurface beam scanners[J]. Scientific Reports, 2022, 12(1): 5912. https://doi.org/doi:10.1038/s41598-022-09874-4.

[6] Lu S, Gao M, Yang Y, et al. Inter-satellite laser communication system based on double Risley prisms beam steering[J]. Applied optics, 2019, 58(27): 7517-7522. https://doi.org/doi:10.1364/AO.58.007517.

[7] Bigler C M, Rovig Z, McDonald J, et al. Holographic amplification of the diffraction angle from optical phase array for optical beam steering[J]. Applied Optics, 2019, 58(34): G241-G247. https://doi.org/doi:10.1364/AO.58.00g241.

[8] Tao X, Cho H, Janabi-Sharifi F. Optical design of a variable view imaging system with the combination of a telecentric scanner and double wedge prisms[J]. Appl. Opt, 2010, 49(2): 239-246. https://doi.org/doi:10.1364/AO.49.000239.

[9] Debbarma K, Sharma S K, Chieh J C S. Beam steering response of X-band triple-mode horn with three element Risley prism[C]. 2022 IEEE International Symposium on Antennas and Propagation and USNC-URSI Radio Science Meeting (AP-S/URSI). IEEE, 2022: 643-644. https://doi.org/doi:10.1109/AP-S/USNC-URSI47032.2022.9886980.

[10] Fan Dapeng, Zhou Yuan, Lu Yafei, et al. Overview of beam pointing control techniques for rotating biprism [J] China Optics , 2013, 6(2): 136-150.(in Chinese). https://doi.org/doi:10.3788/CO.20130602.0136.

[11] Li Lianghai, Liu Xiangnan, Li Xiaoliang Progress and application research of deep space laser communication [J] Journal of Deep Space Exploration, 2019, 6(6): 523-529. (In Chinese). https://doi.org/doi:10.15982/j.issn.2095-7777.2019.06.002.

An Accurate Method for Measuring the Low-Temperature Coefficient of Resistors

Yumei Cui, Hongliang Chen, Ting Ai, Rongwei Feng and Shangming Sang

Beijing Orient Institute of Measurement & Test, Beijing 100086, China

e-mail address: cuiym009@163.com, hbhschl@sina.com, aitiing@163.com, fengrwry@163.com and sangshangming@163.com

Abstract. The temperature coefficient of resistors is one of the key parameters to characterize the reliability of resistors. In precision circuit design, it is necessary to test and screen the key resistors and use the resistors with matching temperature coefficient values to ensure the accuracy and stability of circuit output. In this paper, aiming at how to accurately test the temperature coefficient value of low-temperature coefficient resistors, an indirect test method is proposed. The study adds a reference resistor into the test loop and tests the voltage values at both ends of the reference resistor. The tested resistor when the forward and reverse voltage sources are applied. The accurate resistance values of the tested resistor at different temperatures are obtained by calculation to calculate the temperature coefficient value of the resistor. The method can effectively reduce the measurement uncertainty of the test instrument and the potential thermoelectric influence generated in the test circuit. The measured results show that the measurement method proposed in this paper can obtain a more accurate and reliable resistor temperature coefficient.

1. Introduction

With the progress of science and technology, at present, in the fields of national defense testing, testing, and measurement, the requirements for the accuracy and stability of instruments and meters are higher and higher. Resistor is a necessary and important basic element in designing key circuit modules such as precision amplifier circuits, voltage divider circuits, ADC conversion circuits, and DAC conversion circuits in high-precision instruments. The error caused by the resistor is one of the important factors that affect the output accuracy of this kind of precision circuit module, thus affecting the output accuracy or test accuracy and stability of the whole instrument. The error of the resistor includes the error of resistance value and temperature drift. The error of resistance value is a system error, which can be eliminated by the linear calibration method [1]. Because there is no ideal resistor, all resistors have temperature drift errors. The errors caused by temperature are random errors that cannot be eliminated and can only be minimized [2]. To minimize the error caused by the temperature drift of resistors, it is necessary to test and screen the temperature drift coefficient of resistors [3] and select resistors with matching polarity and size of temperature coefficient to be applied to circuit design to offset part of the error influence caused by temperature drift.

2. Influence of temperature coefficient of resistors

For two resistors with the same nominal temperature coefficient of ±25 ppm/°C, due to different batches and other reasons [4], the actual temperature coefficient of resistors may be +25 ppm/°C or -25 ppm/°C at most. If resistance matching is not carried out during use, when these two resistors are used simultaneously in circuit design, the maximum relative temperature coefficient error may be +50 ppm/°C or -50 ppm/°C, as shown in curves a or b in Figure 1. However, if matching resistors are used according to the application function of the circuit, that is, two resistors with the same temperature coefficient and the same or opposite polarity are used, the relative temperature coefficient error generated in the circuit will be greatly reduced, as shown in curve c (curve d or e) in Figure 1.

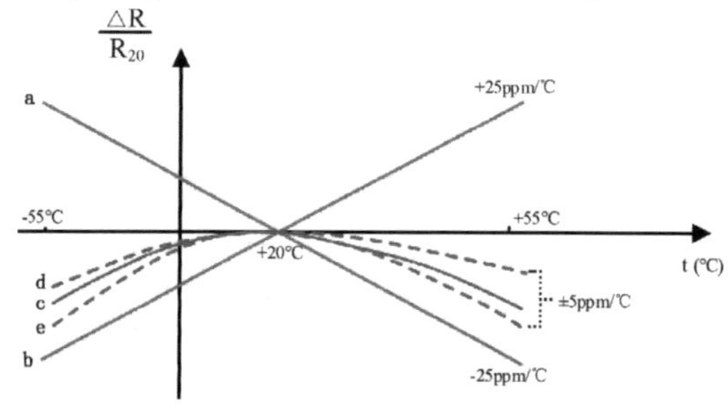

Figure 1. Schematic diagram of resistor matching temperature coefficient

In the amplifier circuit, the temperature drift error mainly comes from the resistor in the circuit. The influence of the resistor temperature coefficient on amplifier gain is analyzed below.

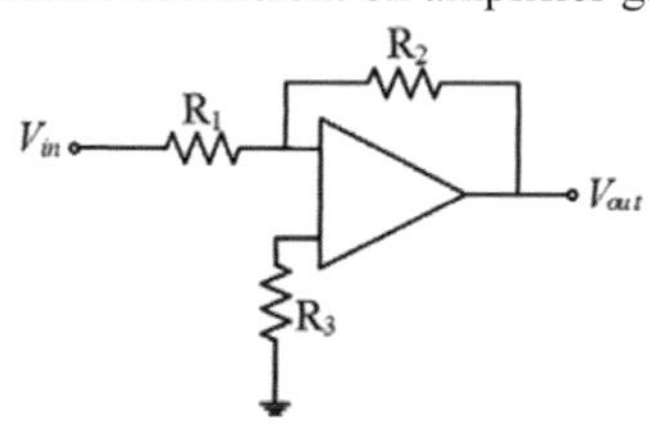

Figure 2. Schematic diagram of inverted operational amplifier

For an inverting input amplifier, as shown in Figure 2, its gain K is:

$$K = -\frac{R_f}{R_i} = -\frac{R_2}{R_1}. \tag{1}$$

The temperature coefficient of the input resistor R_1 is m, and the temperature coefficient of the feedback resistor R_2 is n.

Assuming that the temperature changes by t °C, the gain after the temperature change is:

$$K' = -\frac{R_2+\triangle R_2}{R_1+\triangle R_1} = -\frac{R_2+R_2\times tm}{R_1+R_1\times tn} = -\frac{R_2}{R_1}\times\frac{1+tm}{1+tn}. \tag{2}$$

According to Formula (2), only when the temperature coefficients of the input resistor R_1 and the feedback R_2, that is, the positive and negative polarities of m and n are the same and the numerical values are the same [5], can there be no influence on the gain of the amplifier and the accuracy and stability of the output of the amplifier be guaranteed. If the polarities of the temperature coefficients of the two resistors are opposite or the values are quite different, the temperature drift has a great influence on the accuracy of the amplifier circuit.

From the above, it can be seen that in the inverting amplifier circuit, if the input resistor and the feedback resistor are selected with the same polarity of temperature coefficient, R_1 and R_2 will become larger or smaller in the same proportion with the change of environmental temperature [4], and R_1 and R_2 will offset some error effects caused by temperature drift. Therefore, the temperature coefficient matching of resistors in precision circuits is more important than the absolute temperature coefficient of a single resistor. In fact, it is difficult to make the polarity and value of the temperature coefficient of resistors the same, especially since the temperature coefficient of resistors in different production batches may vary greatly. To ensure the accuracy of precision circuit output, it is necessary to test and screen the temperature coefficient values of key resistors and select resistors with matching temperature coefficients to be applied to the circuit.

3. Test method for temperature coefficient of resistors

At present, the most direct and easy way to test and screen the temperature coefficient of the resistor in the laboratory is to test the resistance value of the resistor under different temperatures directly [6] by using a digital multimeter or a resistance meter, and then calculate the temperature coefficient value [7]. With the improvement of manufacturing quality and precision of resistors, the temperature coefficient of many precision resistors can be very low. The nominal temperature drift coefficient of precision resistors on the market can reach ±(10~50) ppm/°C, even as low as ±5 ppm/°C and ±2 ppm/°C, and the lowest can reach ±0.2 ppm/°C.

When testing the resistance value of a resistor at different temperatures, the measurement uncertainty index of the ohmmeter used is lower than the temperature coefficient of the measured resistor. The measured resistance value will have errors and cannot be used to calculate the temperature coefficient of the resistor. Otherwise, the wrong result will be obtained. So, it is necessary to use a resistance meter with its own measurement accuracy index better than the temperature coefficient of the measured resistor. The 34401A high-precision digital multimeter produced by Agilent Company (later replaced by the German Keysight 34461A model, but its functional indicators remain unchanged) is the most widely used in China [8]. The uncertainty of the resistance test of 34401A or 34461A digital multimeter is 20×10^{-6}. The test error will occur when testing the precision resistance whose nominal temperature coefficient value is equal to or lower than 20 ppm/°C.

In this paper, according to the characteristic of precision resistors with low-temperature coefficients, the accurate test and screening method of temperature coefficient is studied. The test method is to indirectly calculate the temperature coefficient value of the measured resistor by testing the voltage values of the reference resistor and the measured resistor. The schematic test diagram of the indirect test temperature coefficient is shown in Figure 3.

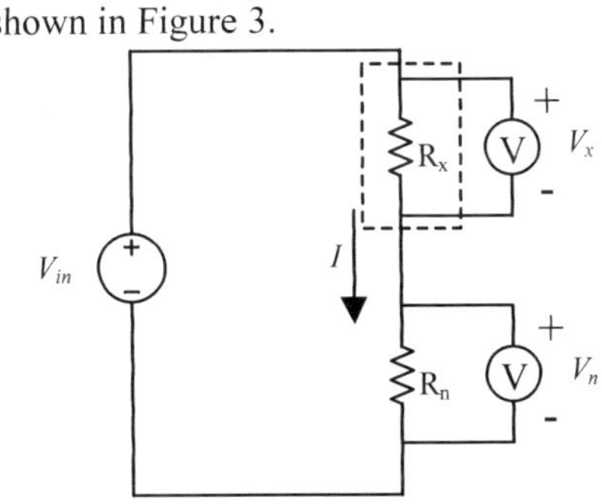

Figure 3. Schematic diagram of indirect test resistor temperature coefficient

V_{in} is the DC voltage source, R_x is the measured resistor, R_n is the reference resistor, and V_1 and V_2 are DC voltmeters.

The test method uses a standard resistor with the same resistance value as the measured resistor, but the temperature coefficient index is better than the measured resistor as a reference resistor. If it is not convenient to find a standard resistor with a better temperature coefficient index, one of the measured resistors can be chosen as a reference resistor. Because the index of the measured precision resistor is relatively high, it has little influence on the test results and accuracy. The reference resistor is connected in series with the resistor to be tested. A DC voltage source is applied to form a test loop. Use two voltmeters of the same model to test the voltage values at both ends of the resistor under test and the reference resistor at the same time.

$$V_x = I \times R_x \tag{3}$$

$$V_n = I \times R_n \tag{4}$$

We calculate Formula (3) and Formula (4), which can get the value of R_x.

$$R_x = -\frac{V_x}{V_n} \times R_n \tag{5}$$

In the actual test, only the measured resistor (that is, the part in the dotted box in Figure 3) is placed in the constant temperature control box. The DC voltage source, the reference resistor, and the voltmeters are all placed in the laboratory environment. The input DC voltage source remains unchanged. Two voltmeters are used to test the voltage values at both ends of the measured resistor at different temperatures and the voltage values of the reference resistor in the laboratory environment.

According to the calculation Formula (5), the resistance values of the resistors under test at different temperatures are obtained indirectly from the voltage values measured by two voltmeters and the resistance values of the reference resistors.

Because the laboratory environment is basically stable, the voltage measurement accuracy and stability of the voltmeters are high. The measured voltage data are stable and reliable, and the error brought by comparison in the calculation formula is extremely small. Because the temperature coefficient of the reference resistor is very low, the resistance value of the reference resistor is basically stable during the test. Therefore, the resistance value of the measured resistor at different temperatures obtained by this indirect test method is accurate.

The temperature coefficient of resistors usually uses the average resistance temperature coefficient [9], which refers to the average relative change of resistance value when the temperature changes by 1 °C in a certain temperature range and is expressed by α.

$$\alpha = \frac{R_2 - R_1}{R_1(t_2 - t_1)} \tag{6}$$

where R_1 and R_2 are the resistance values of the resistor to be measured at temperatures t_1 and t_2 respectively.

4. Accurate test method for eliminating line thermoelectric potential

In order to get a more accurate temperature coefficient of the resistor by test and calculation, the test method and calculation method need to be further optimized.

Since the accuracy of the reference resistor and measured resistor is generally high, the voltage drops [10] on the test line, that is, thermoelectric potential, will affect the accuracy of the test [11]. The thermoelectric potential that affects the test accuracy is mainly generated by the test circuit that tests the voltage values across the resistor, as shown in e_1 and e_2 in Figure 4.

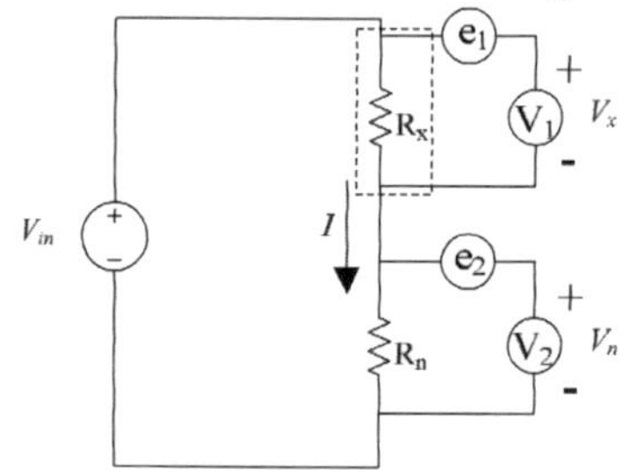

Figure 4. Schematic diagram of thermoelectric potential in the test circuit

In order to reduce the influence of thermoelectric potential produced in the test circuit on the test accuracy, the real ohmic test method of Fluke 8508A is referenced. When the true ohm test is used, the test circuit is connected with the positive measuring current to measure the voltage drop on the resistor. It is then connected with the negative measuring current to measure the voltage drop on the resistor to be measured. The sum of the absolute values of the two measurements is obtained by subtracting the second measurement result from the first measurement result. The sum of the absolute values is divided by 2 to obtain a resistance value that eliminates all the influences of thermoelectric potential [12].

In order to ensure the test accuracy of the temperature coefficient value of the resistor, the above-mentioned true ohm test method is used. When testing the resistance value of the resistor to be tested at a certain temperature, a forward DC voltage source and a reverse DC voltage source of the same value are applied to the test circuit, respectively, as shown in Figures 4 and 5.

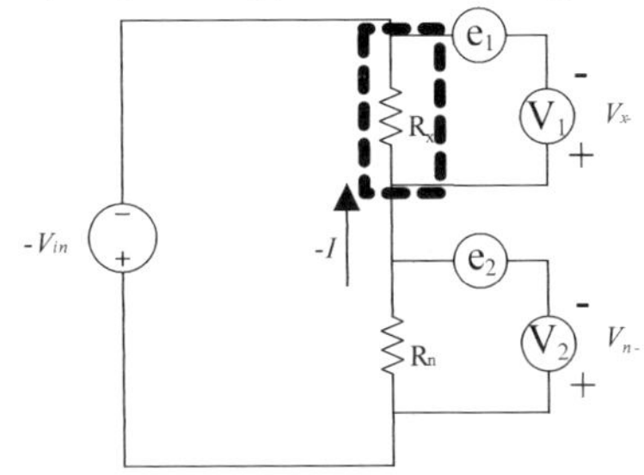

Figure 5. Test schematic diagram of applying the same value of reverse voltage

In the laboratory environment, the thermoelectric potential on the test line is relatively fixed when the ambient temperature is relatively unchanged, and the material of the test line has not changed.

When the forward DC voltage source is applied, the voltages at both ends of the measured resistor and the reference resistor are shown in Formulas (7) and (8), respectively.

$$V_{x+} = I \times R_x + e_1 \tag{7}$$

$$V_{n+} = I \times R_n + e_2 \tag{8}$$

When the same value reverse DC voltage is applied, the voltages of the measured resistor and the reference resistor are as shown in Formulas (9) and (10), respectively.

$$V_{x-} = -I \times R_x + e_1 \tag{9}$$

$$V_{n-} = -I \times R_n + e_2 \tag{10}$$

By subtracting the formulas, i.e. (7) - (9), (8) - (10).

$$V_{x+} - V_{x-} = 2I \times R_x \tag{11}$$

$$V_{n+} - V_{n-} = 2I \times R_n \tag{12}$$

By dividing Formulas (11) and (12), the value of R_x can be obtained.

$$R_x = \frac{V_{x+} - V_{x-}}{V_{n+} - V_{n-}} \times R_n \tag{13}$$

It can be seen from the above that the thermoelectric potential in the test line can be eliminated through calculation to obtain a more accurate resistance value of the measured resistor at a certain temperature, thereby obtaining a more accurate resistance temperature coefficient value.

5. Test verification

The temperature coefficient values of five resistors with a resistance of 12.5 kΩ, a nominal accuracy of 0.1%, and a nominal temperature coefficient of 10 ppm/°C were tested by the test method proposed in this paper. The test system is built according to Figure 3. Any resistor with the same index as the measured resistor is selected as the reference resistor. The DC voltage sources of +10 V and -10 V are provided by the DC voltage source function of FLUKE 5520. The voltages at both ends of the measured resistor and the reference resistor are tested by two DC voltmeter functions of Keysight

34461A. During the test, the resistor to be tested is placed in a constant temperature control box, and the resistance values at 20 °C and 30 °C are tested, respectively. The resistance and temperature coefficient values of five tested resistors at different temperatures are shown in Table 1.

Table 1. Resistance value (kΩ) and temperature coefficient value of the tested resistor (α).

	20 °C	30 °C	α
R1	12.51071	12.51185	9.11×10^{-6}
R2	12.54975	12.54878	-7.73×10^{-6}
R3	12.51129	12.51085	-3.52×10^{-6}
R4	12.51202	12.51169	-2.64×10^{-6}
R5	12.50992	12.50956	-2.87×10^{-6}

According to the test data in Table 1, the resistance value of the low-temperature coefficient resistor can be accurately measured using the test method proposed in this paper. Its accurate resistance coefficient value can be calculated. That is, resistors with matching temperature coefficients can be selected to produce precision instruments and equipment.

6. Conclusion

This paper analyzes the influence of the temperature coefficient of the resistor on the error of the precision circuit, which will affect the stability and accuracy of the output or test of precision instruments. At present, the temperature coefficient values of many precision resistors are getting lower and lower. Still, because the measurement accuracy of the resistance meter itself is not high enough, the accurate temperature coefficient values cannot be obtained by directly testing the resistance values of resistors at different temperatures. Then an indirect test method using a reference resistor and test resistor is designed. This method can effectively reduce the measurement error of the test instrument itself and the error influence of the thermoelectric potential in the measuring circuit on the temperature coefficient value of the test resistor. It can accurately measure the low-temperature coefficient value of the resistor.

References

[1] Xiong Haodong. (2019) Functional Circuit Design of Small Size Six-and-a-Half Digital Multimeter [D]. harbin university of science and technology. https://kns.cnki.net/kcms2/article/abstract?v=3uoqIhG8C475KOm_zrgu4lQARvep2SAkWf ZcByc- RON98J6vxPv10 b3uBUFc8mjhLbYhSsygikDDFBBmQaM7r2rH5XI5Q3hg&uniplatform= NZKPT.

[2] Li Ang. (2017) Research on low-temperature drift technology of DMM measurement circuit [D]. harbin university of science and technology.https://kns.cnki.net/kcms2kcms2/article/abstract?v=3uoqIhG8C475KOm_zrgu4lQARvep2SAkkyu7xrzFWukWIylgpWWcEv k 3iShCK9QejD3jv6n-GZ3VEdHT81pL6QLwwoXURHo9&uniplatform=NZKPT.

[3] Bian Guangqing, Zhang Limin. (2020) The Application of Curve Fitting Method in Temperature Coefficient Test of Resistance [J]. Metrology Science and Technology.2020,4:18-22. https://kns.cnki.net/kcms2/article/abstract?v=3uoqIhG8C44YLTlOAiTRKibYlV5Vjs7i8oRR 1 PAr7RxjuAJk4dHXouDsqi1Go-aP6dydntF9QX5mB8bDAZ- nsI5QIfWxraXn&uniplat form=NZKPT.

[4] EET ESM EDN. (2021). How does the temperature drift of the resistor affect the overall circuit design [DB/OL]. https://mbb.eet-china.com/forum/topic/93247_1_1.html.

[5] Li Xiaoxia. (2001) Effect of Resistance Temperature Coefficient on Amplifier Gain [J]. Techniques of Automation and Applications.2001,4:42-43. https://kns.cnki.net/kcms2/article/abstract?v=3uoqIhG8C44YLTlOAiTRKgchrJ08w1e7AZ-xq6cidhdWEoJyWU2jOHELrF_BwLoCaaTmdjy- qBQDYtqenEpopmE4XVUc6oq9& uniplatform=NZKPT.

[6] Takenaka K, Ozawa A, Shibayama T, et al. (2011). Extremely Low Temperature Coefficient of Resistance in Antiperovskite Mn3Ag1-xCuxN[J]. Applied Physics Letters, 2011, 98(2): 022103-022103-3.
[7] ASTM B84-07 (2019) Standard Test Method for Temperature-Resistance Constants of Alloy Wires for Precision Resistors [S]. https://www.astm.org/b0084-07r19.html.
[8] Jia Huayu, Liu Li, Zhang Jianguo. (2014) Broadband current-mode operational amplifier in pipelined analog-to-digital converter [J]. Optical Precision Engineering, 2014, 22 (10): 2855-2860. https://kns.cnki.net/kcms2/article/abstract?v=3uoqIhG8C44YLTlOAiTRKgchrJ08w1e7M8Tu7YZds8-FFjzvczr1wUY67aUV7Dt4u7ZXBG9SPiDQ4irumusN24tMa3ztkrnz&uniplatform=NZKPT.
[9] China Standards Publishing House. (2006) GB /T 6148—2005 Test method for temperature coefficient of resistance of precision resistance alloys [S].Beijing. https://www.renrendoc.com/paper/240372525.html.
[10] Qian Wenbin. (2012) Development of a DC voltage divider with low temperature coefficient and low thermal potential [J]. Journal of Jiangsu Institute of Education (Natural Sciences).2012,28(5):26-29. https://kns.cnki.net/kcms2/article/abstract?v=3uoqIhG8C44YLTlOAiTRKgchrJ08w1e7fm4X_ 1ttJAn28b06v2UOtqmJSyLmUrKQ4SyX8gKXbwsrzm6L22AXPzw8Kminob5O& uniplatfor m=NZKPT.
[11] Huang Ze, Cui Wende, Zhao Huaye. (2015) Analysis of Effect by Thermoelectric Potential in RTD Temperature Measurement System [J]. Journal of Astronautic Metrology and Measurement.2015,35(2):22-26. https://kns.cnki.net/kcms2/article/abstract?v=3uoqIhG8C44YLTlOAiTRKibYlV5Vjs7ir5D84 hng_y4D11vwp0rrtb1Ht1GYvGadcWBptsTfpaY84SfmmUbKBIhTWvl9yXXk &uniplatform =NZKPT.
[12] Fluke. (2022). 8508A User's Manual [DB/OL]. https://cn.flukecal.com/literature/product-manuals/8508a-reference-multimeter-users-manual.

Simulation validation based on emission coil of three-axis square Helmholtz magnetic source

Qingming Duan[1a], Bofeng Zou[1b], Ruju Zhang[1c], Yuxin Song[1d], Hu Zhao[1e] and Ruipeng Zhang[2f*]

[1]College of Instrumentation and Electrical Engineering, Jilin University, Changchun, China

[2]Laboratory Management Office, Jilin University, Changchun, China

[a]duanqm@jlu.edu.cn, [b]645443276@qq.com, [c]zhangrj2320@163.com, [d]1127900537@qq.com, [e]zhaohu20@mails.jlu.edu.cn, [*f]zhangrp@jlu.edu.cn

Abstract. In electromagnetic detection, the received signal strength is largely determined by the transmitter's signal strength. Therefore, it is necessary to select and design the transmitting coil to meet the requirements of the electromagnetic launching system. To obtain better-transmitting coils, Maxwell is used to establishing square and toroidal coil models for calculation and compare magnetic field intensity and uniformity. To cope with electromagnetic exploration in a complex environment, the electromagnetic field model of the three-axis toroidal coil and square Helmholtz coil is established. The magnetic field simulation results verify that the three-axis square Helmholtz coil has higher emission field intensity under the same excitation condition, which is suitable for electromagnetic exploration.

1. Introduction

With the development of national economic demand and industrialization, resource exploration technology has put forward new requirements [1][2]. As a geophysical method with good effect, electromagnetic prospecting is indispensable in underground exploration and survey [3]. Especially in the field environment, under the influence of various facilities, how to properly radiate and use devices to improve the transmission power is very important [4][5]. In the case of the limited single transmitting coil, the three-axis transmitting coil is made to improve practicability. Compared with the toroidal coil, the square coil is more stable and convenient. The uniformity analysis of the square Helmholtz coil shows that the magnetic field uniformity is good [6][7]. Li et al. designed a square Helmholtz launcher with good results [8]. When the three-axis square Helmholtz coil is adopted, the output power of the transmitting system is improved. Moreover, the output waveform quality is good, which provides a stable and reliable transmitting field source for electromagnetic exploration and better support exploration.

2. Analysis of space magnetic field of current carrying coil

According to the Bio-Savart law, the magnetic field of a straight current-carrying wire at a certain point in space [9] is:

$$B = \frac{\mu I}{4\pi r}(\cos\theta_1 - \cos\theta_2) \tag{1}$$

where θ is the angle between the point and both ends of the wire, μ is the permeability of the medium,

l is the length of the coil side, r is the vertical distance from the point to the straight wire, O is the origin of the coordinates, n is the number of turns of the wire, and $2a$ is the distance between the coils. The magnetic field of the coil at point Q on the central axis is:

$$B_1 = B_{11} + B_{12}. \tag{2}$$

Each section of the coil calculates the magnetic field of the coil. The above formula calculates the magnetic induction intensity in one direction, and the magnetic field at the origin O is calculated as:

$$B_1 = \frac{2\mu_0 n I l^2}{\pi}\left[\frac{1}{[l^2+(a+z)^2]\sqrt{2l^2+(a+z)^2}} + \frac{1}{[l^2+(a-z)^2]\sqrt{2l^2+(a-z)^2}}\right]. \tag{3}$$

It can be concluded from the above formula that when a=0.5445l, the magnetic field uniformity is the best. The current source is controlled by computer software and loaded onto the three-axis square Helmholtz coil. When z=0, the magnetic field intensity is:

$$B_1 = \frac{4\mu_0 n I l^2}{\pi(l^2+a^2)\sqrt{2l^2+a^2}}. \tag{4}$$

3. Simulation of the magnetic source coil

Better electromagnetic emission intensity can be achieved by adopting three-axis square Helmholtz coil emission to achieve deeper detection depth and obtain more geological information [10]. To verify the rationality of the design, the first simulation comparison is made between the toroidal coil and the square coil with a single axis. See Figure 1

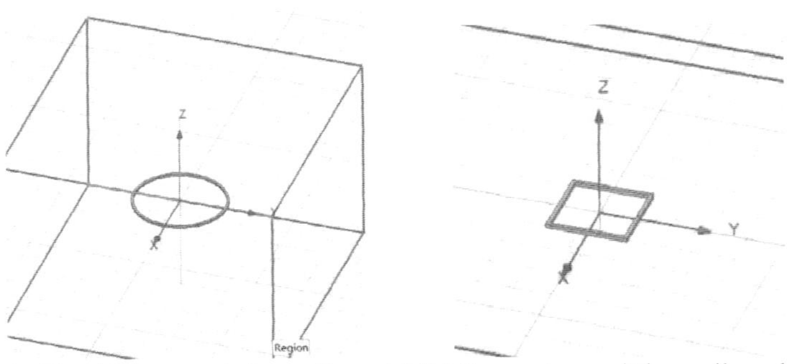

(a) Toroidal transmitting coil model (b) Square transmitting coil model
Figure 1. Model diagram of transmitting coil

The radius of the toroidal coil is 1 m, while the side length of the square transmitting coil is 1 m, and the space area is three times the coil's range. It is wound with copper wire. The simulation space is air to simulate the conditions of the wild environment. The wound side length of the coil is 0.05 m. The same conditions are set up in the simulation test, and the grid is divided to solve the magnetic field intensity. When the excitation current is set to 100 A and the signal frequency is set to 62.5 kHz, the two-coil test is carried out. The magnetic field intensity is shown in Figure 2 and Figure 3.

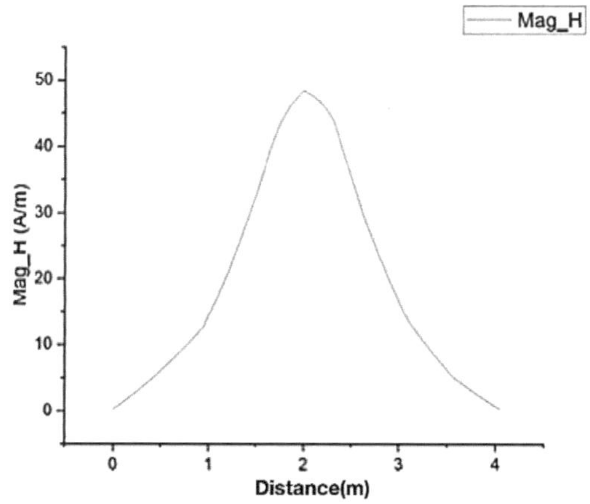

Figure 2. The calculation result of toroidal coil strength

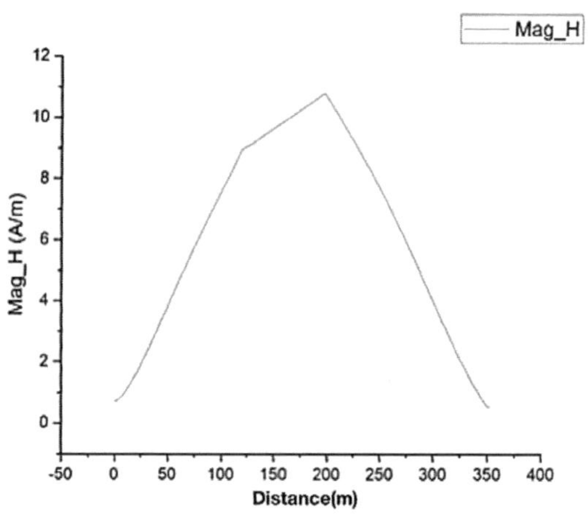

Figure 3. Magnetic field strength calculation diagram of square coil

According to the simulation results of the magnetic field intensity of a single axis under the above setting conditions, the magnetic field intensity of a toroidal coil is better than that of a square coil due to its large area. When converted to a similar area, the magnetic field intensity is close to that of the square coil.

The electromagnetic field analysis of the designed triaxial transmitting coil is needed to prove that its transmitting magnetic field intensity is more excellent. Therefore, the coil model in the case of triaxial emission is constructed and analyzed, as shown in Figure 4. The diameter of the three-axis toroidal coil from inside to outside is 1 m, 1.2 m, and 1.4 m, respectively. The side length of each axis coil of the three-axis square Helmholtz coil is also 1 m, 1.2 m, and 1.4 m. The single turn of the coil is 0.05 m*0.05 m square. Currently, the model of the three-axis square Helmholtz coil and the three-axis toroidal coil is constructed, and the selected parameters are consistent with those set in uniaxial transmission.

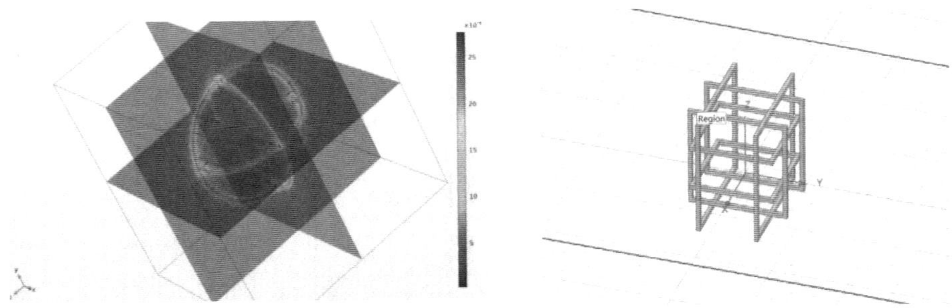

(a) Triaxial toroidal coil model diagram (b) Triaxial square coil model diagram
Figure 4. Three-axis ring model model

Simulation results of the magnetic field strength of three-axis rings are shown in Figure 5 and Figure 6:

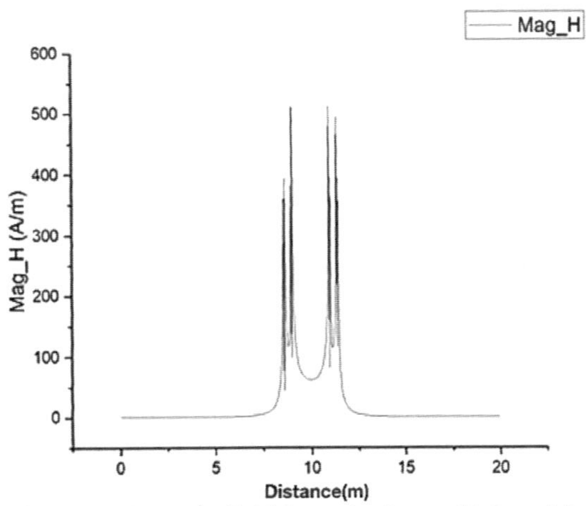

Figure 5. Magnetic field intensity in toroidal model

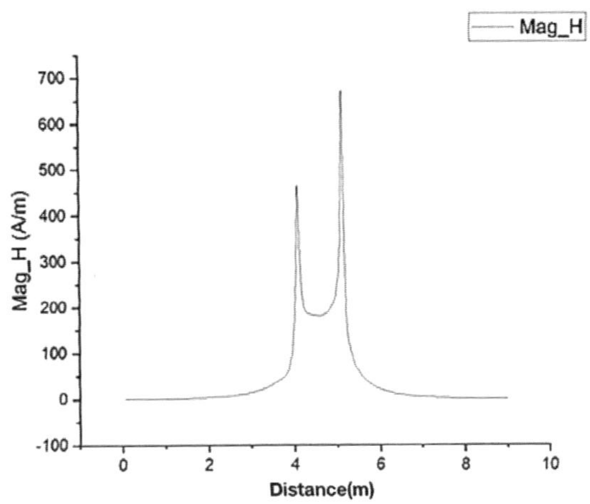

Figure 6. Magnetic field intensity in triaxial square model

The magnetic field results show that under the same emission conditions, the magnetic field intensity emitted by the three-axis square Helmholtz coil is better than that of the three-axis toroidal coil and higher than that of the established single-axis emitting coil.

4. Conclusion

Helmholtz coil works well as a way to generate a uniform magnetic field. Toroidal processing and positioning in the field environment are inconvenient, but the square structure is convenient to make and use. At the same time, the formula analysis shows that the magnetic field uniformity is the best when $d = 0.5445\ l, a = 0.008\ l,$ and $b = 1.077687\ a$. The launch largely determines the reception effect of exploration. Through the theoretical calculation of the toroidal and square coil, it is concluded that the magnetic field intensity of the toroidal coil is slightly better than that of the square coil, but the uniformity is poor. Compared with the calculation results of the three-axis coil simulation model, the three-axis square Helmholtz coil has greater magnetic field intensity, broad application prospect, and strong adaptability.

References

[1] Zhang, S.K., Jiang, P., Lu, L., Wang, S., Wang, H.H. (2022) Evaluation of Compressive Geophysical Prospecting Method for the Identification of the Abandoned Goaf at the Tengzhou Section of China's Mu Shi Expressway. Sustainability, 14: 13785.

[2] Yang, H., Li, S.Q., Liu, G.Q., Li, Y.H., Han, Y.N., Zhang, L.F. (2016) Research on the Effectiveness of Electrical Detection. International Conference on Electrical Engineering and Automation (ICEEA). Xiamen.

[3] Wang, Y. (2020) Research on Unmanned Airborne Transient Electromagnetic Processing Technology in Geological Hazard Exploration. Jilin University, Changchun.

[4] Di, Q.Y., Fu, C.M., An, Z.G., Xu, C., Wang, Y.L., Wang, Z.X. (2017) Field testing of the surface electromagnetic prospecting system. Applied Geophysics, 14: 449-458.

[5] Tripp, A.C. (2005) Acheron's rainbow: Free associations on 75 years of exploration geo-electromagnetics. Geophysics, 70: 25TH.

[6] Li, H., Zhang, Y.S. (2019) Analysis of magnetic field uniformity in triaxial orthogonal Helmholtz coils. Modern Machinery, 1-5.

[7] Tan, X., Liu, J., Yin, J.L. (2012) Magnetic field uniformity of square Helmholtz coils. Optical Instrument, 39-44.

[8] Li, D.L., Sun, X.Y., Hu, T.T., Yang, Y.X., Song, J.Z. (2023) Design of magnetic field generator based on square Helmholtz coil. Journal of Yanshan University, 1-8.

[9] Zhang, S.H. (2009) University Physics. Tsinghua University Press, Beijing.

[10] Geng, S.Y. (2008) Development of Hybrid-source Electromagnetics Transmitter EM-II. Jilin University, Changchun.

Application and Error Analysis of Differential Sampling for AC Power Measurement based on Quantum Voltage

JingFen Bai[1a], Jing Meng[1b], YuBo Yang[1c*], MeiMei Duan[2d]

[1]China Electric Power Research Institute Co., Ltd, Beijing 100192, China

[2]State Grid Jiangsu Electric Power Co., Ltd Marketing Service Center, Nanjing 210019, China

[a]jfbai@epri.sgcc.com.cn, [b]mengj@epri.sgcc.com.cn, [c*]yangyb3333@163.com, [d]cbadmm@163.com

Abstract. The quantum voltage standard based on the Josephson effect has been applied in AC power measurement. Differential sampling is the most promising method for measuring AC signals based on quantum technology. In this paper, we developed an AC power measurement system based on differential sampling and analyzed the system error sources through simulations and experiments. The system error comes from the sampling rate, step number of stepped waves, phase difference, and sampling time. The traceability of power to quantum voltage using differential sampling technology was achieved with an uncertainty of 10^{-6}.

1. Introduction

In the field of measurement, the traditional voltage reference uses the average value of a group of saturated Weston standard batteries. This voltage reference composed of standard batteries is also called the physical reference, with an uncertainty of 10^{-6} [1-3]. However, its characteristics will change slowly with time and the change of temperature. Higher measurement standards are essential for the rapid development of science and technology in industry and agriculture. By the 1980s, the American National Institute of Standards and Technology (NIST) and German Federal Institute of Physical Technology (PTB) had worked together to solve the problems such as the stability of the quantum voltage step generated by the Josephson junction and the microwave distribution. The first Josephson junction array was successfully developed to generate voltage standards. The 10 V DC Josephson voltage standard has been established and widely applied in many major calibration laboratories.

As the DC quantum voltage standard has been widely used, metrologists began to research the AC quantum voltage standard. The two types of AC Josephson voltage standards commonly used are programmable Josephson voltage standard (PJVS) and pulse-driven AC Josephson arbitrary waveform synthesizer (JAWS) [4-6]. At the beginning of the 21st century, the PTB and the NIST successively proposed a differential sampling method, which uses a high-precision digital multi-meter to sample the differential signal between the quantum sinusoidal step voltage and the standard sinusoidal AC voltage. The standard sinusoidal AC voltage signal is reconstructed by eliminating the dynamic sampling points in the quantum step transition process. The measurement uncertainty of AC voltages below 1 kHz reaches 10^{-7} using sampling technology [7,8]. Differential sampling technology has been widely used and has promoted the traceability of AC/DC voltage and power [9-11]. In this paper, we designed an AC power measurement system based on quantum voltage standards using differential sampling to measure voltage and current. We also analyzed the error and measurement uncertainty of the measurement system through simulations and experiments. The feasibility of the differential sampling technology and the accuracy of the measurement system are discussed.

2. Josephson voltage standards

The quantum voltage standards depend on the Josephson effect. In 1962, British scientist Josephson found that when the voltage on both sides of the "Superconductor-Insulating Barrier-Superconductor, SIS" structure is 0, the electrons on both sides of the insulation layer will form Cooper electron pairs [7]. When Cooper electron pairs pass from one superconductor through a very thin insulation layer to another, there is no voltage drop on both ends of the superconductor because the flow of electrons generates the current. Therefore, DC superconducting current is generated at both ends of the array. In this case, both ends of the Josephson junction are in a superconducting non-resistance state. With the increase of current, this non-resistance state is broken, and voltage values begin to appear at both ends of the Josephson junction. This process from the non-resistance state to the resistance state is called the DC Josephson effect. When voltage is applied at both ends of the superconductor, there is an AC superconducting current in addition to the normal DC superconducting current, which causes a voltage drop at both ends of the insulation layer. The frequency of AC current f_0 is proportional to the voltage V_j at both ends, which is called the AC Josephson effect. Josephson effect is the basis for establishing modern quantum voltage reference.

$$f_0 = \frac{2e}{h} \times V_j \tag{1}$$

where e and h are electron charges and the Planck constant. When high-frequency microwave radiation is applied to the Josephson junction, the I-V characteristics exhibit voltage step characteristics, which correspond to the frequencies of microwaves. When a Josephson array is radiated by microwave with frequency f, the Josephson array produces a junction voltage with amplitude V_n.

$$V_n = n\frac{h}{2e} f \cdot N \tag{2}$$

where n is the nth Shapiro step, N is the total number of Josephson knots, and the Josephson constant is defined as $K_j = \frac{2e}{h}$. The constant is suggested by BIPM as a value K_j=493597.85 GHz/V.

2.1. Principle of programmable Josephson Voltage Standard (PJVS)

The programmable Josephson Voltage Standard synthesizes a voltage step-like AC waveform by varying the number of Josephson junctions in operation. Two new junction arrays have been developed, and the new array can quickly select work steps with higher noise resistance and shorter synthesis time. The Josephson junction of PJVS comprises several Josephson subarrays with binary distribution integrated on a single microwave transmission line [9]. The number of Josephson junctions in these subarrays is 1, 2, 4, 8, and 16, ... Each small block has a corresponding offset circuit to offset the Josephson junction to the middle of the step. The whole system is similar to a digital-to-analog converter, which synthesizes a quadratic step-like AC quantum voltage signal by changing the number of Josephson junctions in operation. The structure of the PJVS system is shown in Figure 1.

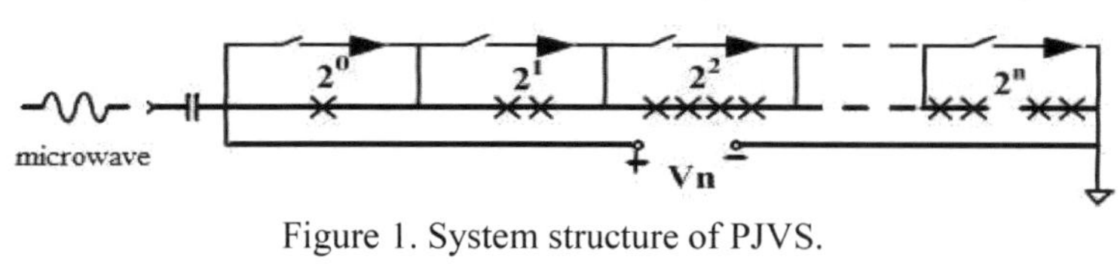

Figure 1. System structure of PJVS.

2.2. Principle of Pulse Driven AC Josephson Voltage Standard (ACJVS)

The pulse-driven AC Josephson Voltage Standard (ACJVS) generates the quantum signal of the synthesized by changing the microwave frequency radiated on the Josephson junction array [12,13]. The pulse-driven quantum voltage reference is based on the Delta-Sigma Modulated analog-to-digital converter. The signal generated by the modulation drives the Josephson array to generate an accurate and pure spectrum voltage waveform, as shown in Figure 2. Similar to the PJVS system, ACJVS needs to connect many array arrays in series to generate accurate quantum voltage. Unlike the PJVS, ACJVS uses the current pulse generated by the pulse generator to bias the array. So, in each clock cycle, each junction in the array will be driven by a positive pulse, a negative pulse, or a zero pulse.

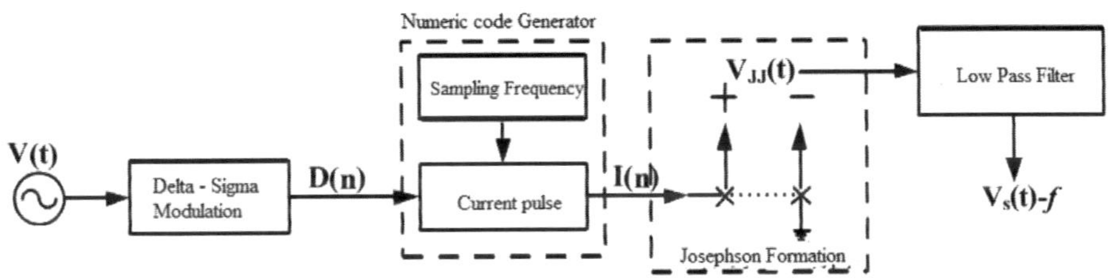

Figure 2. Composite schematic diagram of ACJVS.

The PJVS array has a more stable performance. The synthesized voltage signal can be used not only as a DC voltage standard but also as an AC voltage standard within 1 kHz and has been used in many AC measurement fields [12-15]. PJVS is more resistant to current noise and can generate relatively accurate quantized voltage. However, there are unavoidable differences between the approximate waveform of the synthesized step voltage and the expected synthesized sinusoidal AC voltage waveform, resulting in more high-order harmonic components in its spectrum. The voltage step generated by the programmable Josephson junction array is switchable, and the effective value of the signal is closely related to the switching time. It is difficult to determine the output value of the array voltage during the switching process.

The voltage waveform synthesized by the ACJVS system is close to the ideal sine wave. The accuracy of the synthesized voltage is high, the frequency spectrum of the voltage signal is very pure, and the frequency coverage ranges from 150 Hz to 100 kHz [13-15]. However, compared with the PJVS system, the microwave pulse bias of the ACJVS system is more complex, and the number of corresponding junctions is much less than the PJVS system, leading to a decrease in the amplitude value of the synthetic voltage.

3. Differential sampling and design of power measurement system

According to the principle of differential sampling [16,17], we designed a power measurement system based on quantum voltage standards, as shown in Figure 3. The power measurement system includes a power source, a quantum voltage system, a differential sampling system, a voltage amplifier, and a voltage-to-current conversion device. The power source generates a test voltage signal V_V and a voltage signal V_I that reflects the measured current signal. The phase of the voltage signal is adjustable. After the voltage signal V_V passes through the voltage amplifier, a voltage signal U with an effective value of 120 V is obtained. After the voltage signal is amplified by a converter and transconductance amplifier, a current signal I with an effective value of 5 A is obtained. The amplified voltage and current signals are input to the device under test to complete the power measurement.

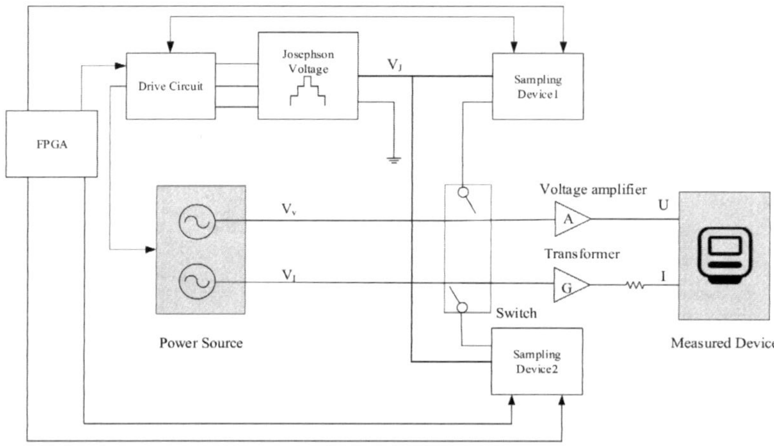

Figure 3. AC power measurement system based on differential sampling.

To improve the accuracy of precise measurement of quantum voltage, it is necessary to ensure that the amplitude of the AC quantum voltage waveform and the measured AC waveform are similar and that the two intersect in the middle of the AC quantum voltage step to minimize the amplitude of the differential signal, achieving the ideal differential sampling. To make the phase and amplitude of V_V or V_I consistent with that of V_J, a differential sampling and control system was designed. A differential sampling control method based on discrete Fourier transform was designed to precisely measure amplitude and phase. The principle is shown in Figure 4.

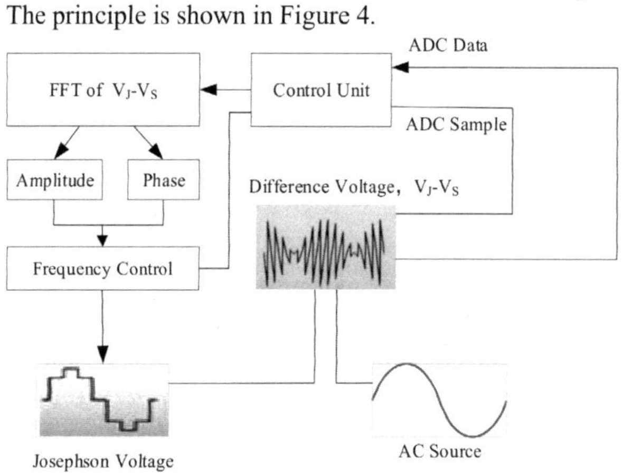

Figure 4. Differential sampling control system based on DFT.

3.1. Error analysis of sampling system

The differential signal and the recovered sine voltage signal are shown in Equation (3):

$$\begin{cases} U_{ac} = u_0 \cdot \sin(2\pi \frac{n}{N} + \Delta\varphi), n = 0,1,\ldots,N-1 \\ U_j = u_j \cdot \sin(\frac{2\pi}{M} + Int(\frac{n}{N/M})) \\ \Delta U = U_{ac} - U_{pj} \\ U^{'} = \Delta U + U_j \end{cases} \quad . \tag{3}$$

In Equation (3), U_{ac} is the ideal AC sine voltage sampling signal. U_j is the step wave that approximates the sine wave, U' is the reconstructed signal after differential sampling, $\Delta\varphi$ is the phase difference between the sine wave U_{ac} and the step wave U_j. N is the number of sampling points per cycle, which is positively correlated with the sampling rate, M is the number of sampling steps, n is the sequence number of differential sampling, and $Int(\frac{n}{N/M})$ represents the rounding operation on $\frac{n}{N/M}$.

When the voltage step is flat, the sampled data is effective. The sine signal obtained by sampling can be expressed by Equation (4):

$$U_{ac} = u_0 \cdot \sin(2\pi \frac{j}{N} + \varphi_1), j = [n\frac{N}{M} + s, n\frac{N}{M} + t - 1]. \tag{4}$$

In Equation (4), s and t represent the sampling point at all times.

Therefore, the selection of sampling points N, sampling steps M, phase difference $\Delta\varphi$, and sampling timing will all affect measurement errors.

Choosing the appropriate number of steps for the step wave can improve measurement accuracy. As the number of steps increases, more sampling points are dispersed from the sine signal, and the error will also decrease. However, as the number of steps increases, the proportion of the transition process to the step level will increase. Therefore, it is necessary to choose the appropriate number of steps according to actual needs. When the signal f =50 Hz, the sampling rate is f_s = 80 kHz, and the number of steps M of the AC quantum voltage is changed, the change in error is shown in Table 1.

Table 1. Effect of steps number on measurement accuracy.

Steps number *M*	**10**	**20**	**40**	**80**
Amplitude/V	0.9826061	0.9963524	0.9989179	0.9997412
Error/V	-0.0173939	-0.0036476	-0.0010821	-0.0002588

The selection of sampling rate also affects measurement accuracy. When the number of steps remains constant, increasing the sampling rate increases the number of sampling points on each step. However, due to the influence of noise interference, the error no longer decreases when the sampling rate is increased to a certain extent. When the signal f =50 Hz, the number of steps $M = 40$, and the sampling rate f_s of the AC quantum voltage is changed. The change in error is shown in Table 2.

Table 2. Effect of sampling rate on measurement accuracy.

Sampling rate f_s/kHz	**10**	**20**	**40**	**80**
Amplitude/V	0.9836061	0.9983528	0.9983256	0.9984970
Error/V	-0.0163939	-0.0016472	-0.0016744	-0.0015030

3.2. Error analysis of voltage and current converters

The ratio of voltage to current is an important link in power measurement. In this paper, a 100:1 resistive voltage divider is used to convert a 1 V AC voltage to a 120 V voltage signal. The 1 V voltage signal is converted to a 5 A current signal by using a 10-ohm resistor shunt and a 0.5 A/5 A current transformer. The error of a resistive voltage divider comes from the phase error at different frequencies, and a commercially available resistive voltage divider can reach -2.5 μrad at 50 Hz. The error of the current transformer comes from proportional error and phase error, and its proportional error can reach 1.5 μA/A at 50Hz. The phase error is 0.6 μrad, and the error of the resistance is 0.3 μΩ/Ω.

A hardware verification platform has been established in our laboratory. Fluke 5720A is selected as a signal source due to its adjustable frequency and amplitude, low noise, high accuracy, and excellent nonlinear performance. The 8-and-a-half-bit digital multimeter 3458A is selected as a differential sampling device. NI-6733 is selected as a sinusoidal step wave generator for outputting AC step waveforms. The entire system achieves clock synchronization using FPGA.

During the test, the peak value of AC sinusoidal voltage and quantum voltage is set to 1 V, the frequency is set to 50 Hz, and the number of steps per waveform cycle is set to N=64. Different amplitudes and phases are set, and the test results are shown in Table 3. The amplitude error is better than ± 0.5%, and the phase error is better than ±0.2°. Due to signal source fluctuations, asynchronism, and algorithm boundary conditions, there are certain differences between the measured data and the simulation results. When the relative amplitude and phase are small, the error is relatively large. Based on the above analysis, the power measurement uncertainty is calculated from Equations (5) and (6), and the power measurement uncertainty is 9.5 μW/VA (k = 2).

Table 3. Uncertainty analysis.

Standard uncertainty	**Source**	**Uncertainty component /10^{-6}**
u_1	Measured device	2.0
u_2	Differential sampling system	0.2
u_3	Inductive voltage divider	3.1
u_4	I-U conversion	3.5

$$u^2 = u_1^2 + u_2^2 + u_3^2 + u_4^2 \tag{5}$$

$$\begin{cases} u_1 = \cos\varphi \cdot u(\dfrac{\Delta U}{U}) \\ u_2 = \cos\varphi \cdot u(\dfrac{\Delta I}{I}) \\ u_3 = \sin\varphi \cdot u(\Delta\varphi_U) \\ u_4 = \sin\varphi \cdot u(\Delta\varphi_I) \end{cases} \tag{6}$$

4. Conclusion

An AC power measurement system based on quantum voltage is developed in this paper. The voltage and current are measured by differential sampling. The step wave and signal source are synchronized based on DFT. The error sources of the measurement system are analyzed, and a hardware verification platform is built to verify the feasibility of the differential sampling technology. The measurement uncertainty of the measurement system reaches the level of 10^{-6}.

Acknowledgments

This work was financially supported by the science and technology project of the State Grid Corporation of China. The project is 'Research on new quantum traceability theory and simulation technology based on quantum technology' (5700-202018170A-0-0-00).

References

[1] Alain, R., Flowers-Jacobs, N. E., & Benz, S. P. (2018). Impact of the latest generation of josephson voltage standards in ac and dc electric metrology. Metrologia, 55(5), S152-S173.
[2] Kieler, O. F., Behr, R., Wendisch, R., Bauer, S., Palafox, L., & Kohlmann, J. (2014). Towards a 1 v josephson arbitrary waveform synthesizer. IEEE Transactions on Applied Superconductivity, 25(3), 1-5.
[3] Underwood, J. M. (2018). Uncertainty analysis for ac-dc difference measurements with the ac josephson voltage standard. Metrologia, 56(1).
[4] Budovsky, I., Georgakopoulos, D., & Hagen, T. (2010). System for precision ac-dc difference measurements based on a Programmable Josephson Voltage Standard. Precision Electromagnetic Measurements. IEEE.
[5] Zhou K L, Qu J F, Zhang Z H, (2017). The status of AC quantum voltage standards [J]. Acta Metrologica Sinica, 38(4): 486 — 491.
[6] Damir Ili, Heinrich, A., Meisner, J., & Behr, R. (2022). Calibration of a precision current measurement system for high ac voltages using an ac quantum voltmeter.
[7] Lee, J., Behr, R., Schumacher, B., Palafox, L., & Fleischmann, P. M. (2016). From AC quantum voltmeter to quantum calibrator. Precision Electromagnetic Measurements. IEEE.
[8] Brevik, J. A., Flowers-Jacobs, N. E., Fox, A. E., Golden, E. B., Dresselhaus, P. D., & Benz, S. P. (2017). Josephson arbitrary waveform synthesis with multilevel pulse biasing. IEEE Transactions on Applied Superconductivity, PP (3), 1-1.
[9] Behr, R., Palafox, L., Ramm, G., Moser, H., & Melcher, J. (2007). Direct comparison of josephson waveforms using an ac quantum voltmeter. IEEE Transactions on Instrumentation and Measurement, 56, p.235-238.
[10] Lee, J., Behr, R., Palafox, L., Alex, Katkov, E., & Schubert, M., et al. (2013). An ac quantum voltmeter based on a 10-v programmable josephson array. Metrologia.
[11] Starkloff, M., Schubert, M., Lee, J., Behr, R., & Peiselt, K. (2016). Dry cooling of a 10 V programmable Josephson voltage standard array. Precision Electromagnetic Measurements. IEEE.
[12] Shiv, Kumar, JaiswalY., & Tang. (2009). Direct comparison between the nist 10 v compact josephson voltage standard and the 2.5 v programmable josephson voltage standard. Mapan.
[13] Sosso, A., Durandetto, P., Trinchera, B., Fretto, M., Monticone, E., & Lacquaniti, V. (2016). Cryogen-free operation of snis for ac quantum voltage standards. IEEE Transactions on Applied Superconductivity, 26(3), 1-3.
[14] Motes, K. R., Mann, R. L., Olson, J. P., Studer, N. M., Bergeron, E. A., & Gilchrist, A., et al. (2016). Applications in quantum metrology. Wiley-VCH Verlag GmbH & Co. KGaA.
[15] Rufenacht, A., Howe, L. /., Fox, A. E., Schwall, R. E., Dresselhaus, P. D., & Burroughs, C. J., et al. (2015). Cryocooled 10 v programmable josephson voltage standard. IEEE Transactions on Instrumentation & Measurement, 64(6), 1477-1482.
[16] Waltrip B C, Gong B, Nelson T L, et al. AC Power Standard Using a Programmable Josephson Voltage Standard[J]. IEEE Transactions on Instrumentation & Measurement, 2009, 58(4):1041-1048.
[17] Palafox L, Behr R, Ihlenfeld W, et al. The Josephson effect based primary AC power standard at PTB: Progress report[C]// Conference on Precision Electromagnetic Measurements Digest. IEEE, 2009.

A Novel Pulse Excitation Circuit Model and Analysis of Its Magnetic Induction Advantages

Zhi Chao Li*[1,2], Chun Guang Zhang[1,2], Jia Rui Liu[1,2], Shuai Chu[1,2], Ying Jiao Gong[1,2], Wei Geng[1,2]

[1] R&D center, National Engineering Research Center of Transducer, Beihai Street 242, Shenyang, CHINA
[2] R&D center, SINOMACH Sensing Technology Co., Ltd., Beihai Street 242, Shenyang, CHINA

Corresponding Author's E-mail: 15140211831@163.com

Abstract. In this paper, we discuss the excitation module in the magnetic induction system. First, we start from the traditional sinusoidal input resonant excitation circuit and give the frequency domain equation of its excitation current. Then, we propose a new module based on pulse excitation and give the corresponding excitation equation in the frequency domain. After that, we deduce the mathematical model of excitation gain of new and old modules using application and technology and finally obtain intuitive mathematical analysis graphics for engineering promotion.

1. Introduction

In recent years, Magnetic Induction has been considered a promising solution for wireless communications in underground and underwater environments [1-2]. Wireless magnetic communication is now widely applied in the short-range communication system. As the magnetic field has similar permeability in both air and dense media, it can provide a more stable channel environment for communication and has been increasingly applied in various areas such as underground sensor networks, underwater robots, bio-implantable devices, etc. [7-10].

For Magnetic Induction (MI) wireless communications, signals are transmitted by modulating the magnetic near field of a coil antenna. Research on MI-based communications in challenged and harsh environments [3,4], such as soil, oil reservoirs, and water pipelines, has demonstrated many promising features of MI-based communication. Extremely low frequency (ELF) electromagnetic signal has good penetrability, which can penetrate the corresponding medium to reach the ground. Relying on ELF electromagnetic signals can overcome the disadvantages of conventional acoustic, optical, and electrical signals being shielded by metal pipes, so it is an effective means of positioning and communication.

The traditional excitation process is described as follows: the ELF electromagnetic signal transmitting coil in the medium continuously transmits an ELF electromagnetic pulse with a certain duty ratio and forms an alternating ELF electromagnetic field outside the medium. The current excitation system is generally composed of a sine signal source and finite-length solenoid coil in series [5], and at a fixed frequency point, it can provide a high amplitude carrier signal for wireless sensor modules [6]. However, from other frequency points on the spectrum, the excitation current amplitude has been greatly attenuated, which brings limitations to the research of location technology in complex media environments [11]. In this case, it is sometimes difficult to use other frequency bands when positioning the in-pipe radiation detection robot [12]. Similarly, the process of transmitting underground sensing information to the above-ground system generally requires routing to achieve reliable frequency selection, because the links distributed in different locations have different optimal operating frequencies with MI channels that can achieve maximum gain [13-14].

Although the routing frequency selection mode can only be used in complex media and can communicate flexibly, it cannot analyze the media environment from a macro perspective, such as the internal cracks and plugging in the domestic pipelines, engine pipelines, underground gas pipelines, and sewage drainage pipelines [15]. A soil path attenuation model has been established based on WUSN [18], and the optimal working frequency range and coil adaptive scheme have been determined, forming an efficient system in terms of energy and application bandwidth (BW), which has certain adaptability to the changes of soil parameters. In terms of power analysis and improvement research of the MI system, the induced voltage in the receiving coil is directly analyzed [16] under the condition of constant system power, and the maximum transmission range of the SNR model is deduced under different noise levels. In [17], it focuses on significantly expanding the M2I communication range by using a spherical coil array. In addition, using a sine signal source as an excitation source can simplify the analysis process, but when the communication frequency band is in the ELF range, the sine excitation mode will be significantly affected. After all, ELF electromagnetic wave has a long period, which does not contribute to the design of communication protocol.

To solve the difficulty of a single frequency location in complex media, a novel pulse excitation circuit without resonance principle is proposed, and its mathematical model is established. Then compare the model with the traditional resonant excitation model to obtain the mathematical model of gain, and display the results in the form of a three-dimensional surface.

2. Excitation Circuit MODEL

2.1. Traditional resonant excitation circuit

The model of the typical resonant excitation circuit is shown in Figure 1,

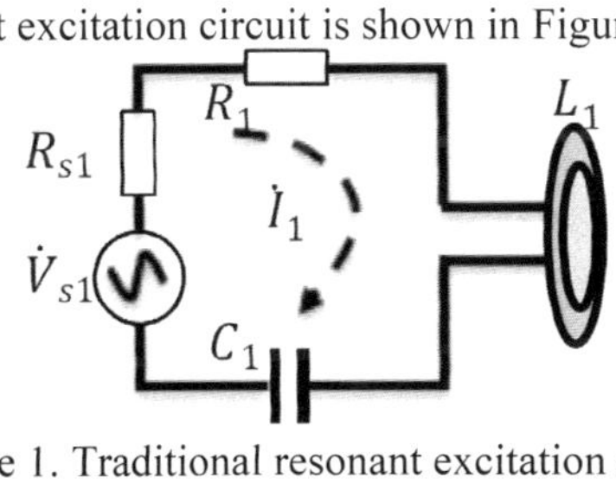

Figure 1. Traditional resonant excitation circuit.

Where $\dot{V}_{s1}$ is the equivalent ac voltage of the power source, R_{s1} is the internal resistance of the power source, R_1 is the internal resistance of coils, L_1 is the resonant capacitance, $\dot{I}_1$ is the excitation current vector. According to Kirchhoff Law, the loop equation is given as

$$\dot{I}_1 = \frac{\dot{V}_{s1}}{R_1 + R_{s1} + j\left(\omega L_1 - \frac{1}{\omega C_1}\right)} \tag{1}$$

The resonant frequency of the coil is obtained by [19]

$$\omega_0 = \frac{1}{\sqrt{L_1 C_1}} \tag{2}$$

In the current near-field positioning research, the resonant frequency of L_1 and C_1 is usually used as the excitation frequency, as shown in Equation 2. Narrow-band electromagnetic wave transmission and reception based on resonant frequency points can meet the positioning requirements in uniform media. However, when the media is composed of metal pipes, rock, soil, and even seawater, it is difficult to accurately locate through a single frequency. To further analyze the frequency characteristics of the circuit in Figure 1, the Laplace transform is applied to Equation 1, and Equation 3 is obtained

$$I_1(s) = \frac{V_{s1}(s)}{R_1 + R_{s1} + sL_1 + \frac{1}{sC_1}} \tag{3}$$

At present, excitation source is generally sinusoidal signal source, and its angular frequency is assumed to be ω_0. After Laplace transformation, Equation 4 is obtained

$$V_{s1}(s) = V_{rms} \cdot \frac{\omega_0}{s^2 + \omega_0^2} \tag{4}$$

Combine Equation 3 and Equation 4, after sorting we have

$$I_1(s) = \frac{V_{rms} \cdot \omega_0 C_1 \cdot s}{s^4 L_1 C_1 + s^3 R_1 C_1 + s^2(\omega_0 L_1 C_1 + 1) + s\omega_0^2 R_1 C_1 + \omega_0^2} \tag{5}$$

Let $s = j\omega$, the frequency response expression of Equation 5 is obtained after sorting

$$I_1(\omega) = \frac{V_{rms} \cdot \omega_0 C_1 \omega}{\omega R_1 C_1 (\omega^2 - \omega_0^2) + j(\omega^2 L_1 C_1 - 1)(\omega^2 - \omega_0^2)} \tag{6}$$

According to Equation 6, when the angular frequency ω be equal to the fundamental frequency ω_0, the equivalent impedance of the circuit is equal to R_1. For the convenience of comparison in the following text, $I_{1m} = \frac{V_{rms}}{R_1}$ is obtained. To facilitate the analysis of other frequencies relative to ω_0, we introduce the frequency shift coefficient $k_\omega = \frac{\omega}{\omega_0}$; when $k_\omega > 3$, $I_1(\omega)$ can be expressed as

$$I_1(\omega) = \frac{V_{rms} \cdot C_1 k_\omega}{k_\omega^2 - 1} \cdot \frac{1}{\omega R_1 C_1 + j(k_\omega^2 - 1)} \cong \frac{V_{rms} \cdot C_1}{k_\omega} \cdot \frac{1}{\omega R_1 C_1 + jk_\omega^2} \tag{7}$$

When Equation 2 and Equation 7 are combined, Equation 8 is obtained, where $X_L = \omega \cdot L_1$

$$I_1(\omega) = \frac{V_{rms}}{k_\omega X_L \cdot \left(\frac{R_1}{L_1} + j\omega\right)} \tag{8}$$

2.2. Novel Pulse Excitation Circuit

The excitation circuit proposed in this paper is analyzed as follows:

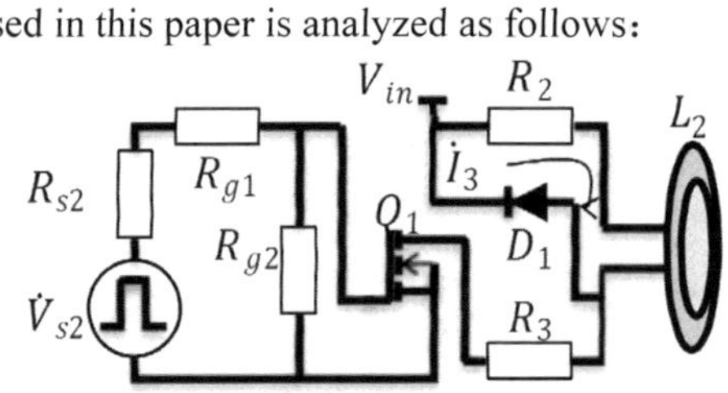

Figure 2. A new type of pulse excitation circuit.

The circuit includes two processes, namely, charging and discharging of the inductor L_2. To facilitate the compatibility of the controller output interface, a positive pulse is used as an excitation control signal, and the inductor is charged to the saturation value of I_3^{max} at a lower current through the current limiting resistor R_3, and then discharged through the internal resistance R_2 of the inductor at a relatively fast speed, wherein $\beta = \frac{R_2}{L_2}$

$$I_3(t) = I_3^{max} \cdot e^{-\beta t} \tag{9}$$

In the frequency domain, Equation 9 is expressed as follows

$$I_3(\omega) = \frac{V_{in}}{(R_2 + R_3) \cdot \left(\frac{R_2}{L_2} + j\omega\right)} \tag{10}$$

3. Comparative analysis of excitation current

Assume that the parameters of inductance in the two circuit models we compare are completely consistent, that is, $R_1 = R_2$, and $L_1 = L_2$, then we only need to care about the left factor of Equation 8 and Equation 10. To further compare the excitation current characteristics of the two circuit models, the excitation gain $G(\omega)$ is introduced. Next, the relationship between inductance $L_{1(2)}$and internal resistance $R_{1(2)}$under different structures is obtained from the inductance processing technology.

To decouple the inductance process parameters, the average diameter $\widetilde{D}$ is used to simplify the gradient diameter of the inductance coil, N represents the total number of turns on the solenoid with length l_N. ρ is the winding resistivity, A_{wire} is the conductive sectional area of the enameled wire, R_{coil} is the DC resistance of the inductive coil, namely the internal resistance $R_{1(2)}$, J is the current carrying capacity of the enameled wire at the corresponding temperature. For the sake of reliability, I_3^{max} is taken as the calculation basis of A_{wire}. Finally, the engineering expression of R_{coil} is obtained through Equations 11~14

$$l_N \doteqdot N\pi\widetilde{D} \tag{11}$$

$$R_{coil} = \frac{\rho l_N}{A_{wire}} \tag{12}$$

$$I_{3m} = JA_{wire} \tag{13}$$

$$R_{coil} = \frac{\rho N\pi\widetilde{D}J}{I_3^{max}} \tag{14}$$

In the previous research on the external magnetic field of the solenoid with multiple coils energized, the arbitrary section of the solenoid is usually taken as the analysis object, and then the integration is carried out in the axial direction. The number of turns per unit section is expressed as follows.

$$n = \frac{N}{l_{sol}} \tag{15}$$

Where l_{sol} is the length of the inductive solenoid, and Equation 14 can be deformed as

$$R_{coil} = \rho\pi J l_{sol} \cdot \left(\frac{n\widetilde{D}}{I_3^{max}}\right) \tag{16}$$

The factor of Equation 10 can be rewritten as follows, where h_{lmt} is the current limiting factor, thus simplifying the calculation of $R_2 + R_3$. Let DC V_{in} be approximately equivalent to V_{rms}.

Assuming that the effective values of the excitation current of the two excitation circuits are equal, then $h_{lmt}=1$

$$\frac{V_{in}}{R_2 + R_3} = \frac{V_{rms}}{h_{lmt} \cdot \rho\pi J l_{sol} \cdot \left(\frac{n\widetilde{D}}{I_3^{max}}\right)} \tag{17}$$

4. Results and analysis

The excitation inductance L_{coil} and inductive reactance X_L can be calculated by the following equation, where μ Is the permeability of the solenoid magnetic core, k_m is the inductance process coefficient, which is inversely proportional to $\widetilde{D}/l_{sol}$, and the value range is 0~1

$$L_{coil} \cong \frac{k_m \mu N^2 \widetilde{D}}{l_{sol}} \tag{18}$$

$$X_L = k_m \mu l_{sol} n^2 \widetilde{D} \omega \tag{19}$$

Combining Equations 8, 10, and 18, the excitation current gain of the new excitation circuit proposed in this paper can be obtained compared with the habitual excitation circuit in the frequency domain.

$$G(\omega) = \left(\frac{k_m \mu n}{\rho \pi J}\right) \cdot I_3^{max} \cdot k_\omega \omega \tag{20}$$

Considering that k_m is inversely proportional to the square root of $\widetilde{D}/l_{sol}$, the expression is as follows,

$$k_m \cong \sqrt{\frac{l_{sol}}{10 \cdot \widetilde{D}}} \tag{21}$$

Equation 20 gives the excitation comparison of the two excitation methods under multiple parameters. To further study the influence weight of each parameter and guide subsequent experiments and winding processing, the excitation winding is made of copper enameled wire, and the winding magnetic core is made of permalloy wire. $k_m \mu n/\rho \pi J$ is linked to the inductance process design. I_3^{max} is the peak value of the excitation current, and $k_\omega \omega$ is a relative quantity, which represents the advantages of the modern excitation circuit model when ω gradually increases from ω_0. Finally, the process parameters of the excitation winding are given in Table 1.

Table 1. Excitation winding process and engineering application parameter table

Parameters	Value	Unit
Current carrying capacity of copper enameled wire (J)	2.5	A/mm²
Resistivity of copper enameled wire (ρ)	0.0172	Ω·m/mm²
Permeability of permalloy wire (μ)	0.125	H/m
Number of windings turns per unit radial length (n)	100	/mm
Inductance process coefficient (k_m)	0.1~1	-
Frequency range (ω)	0.1~25	rad/s
Frequency multiple (k_ω)	1~80	-
Excitation current peak value (I_3^{max})	0.1~1	A

Equation 20 analyzes the gain of the new excitation current relative to the traditional excitation current in a sweep mode, and the overall excitation gain is expressed as $20lg|G(\omega)|$. Where the CT gain coefficient is equal to the peak value of the excitation current I_3^{max}, and the FS gain coefficient is equal to the product of the electromagnetic wave angular frequency ω and the swept frequency range k_ω. According to the parameter setting range in Table 1, the numerical curve in Figures 1-3 is obtained.

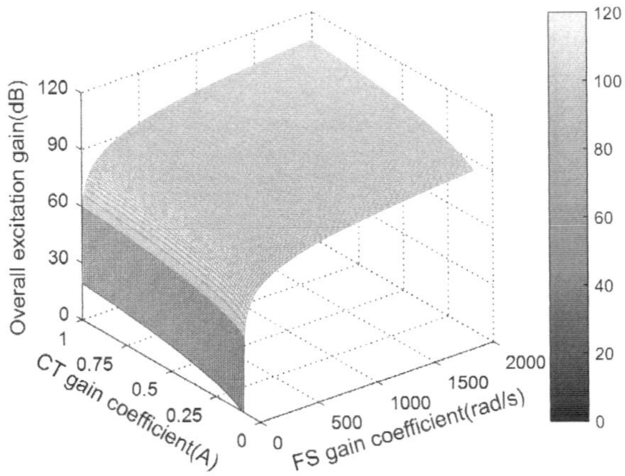

Figure 3. The total excitation gain is displayed when k_m is 1.0.

As can be seen from Figure 3, when the upper limit of the overall excitation gain reaches the maximum value at k_m=1 (i.e., l_{sol}=10 · $\widetilde{D}$), the current gain coefficient and frequency shift factor are allowed to take values within a wide range in engineering applications. Therefore, when there is no special explanation for the size requirements of the excitation winding, the condition k_m=1 should be met as much as possible.

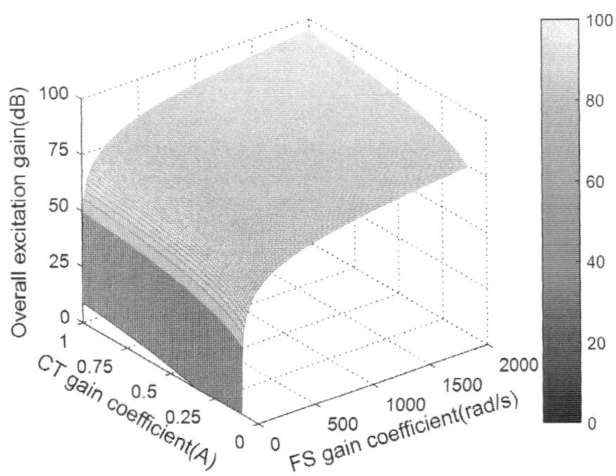

Figure 4. The total excitation gain is displayed when k_m is 0.3.

When the excitation current is lower than 0.3A, or the frequency shift gain is lower than 100 rad/s, it can be observed from the starting point state of the current factor and the frequency shift factor, although the overall excitation gain peak is still very ideal k_m=0.3 (i.e., $l_{sol} \cong \widetilde{D}$), that the overall excitation gain decreases instead. Therefore, using such structure excitation winding, in engineering applications, to obtain the optimal excitation gain, the excitation current should be greater than 0.5A, the frequency shift factor should not be lower than 250 rad/s, and the numerical analysis results are shown in Figure 4.

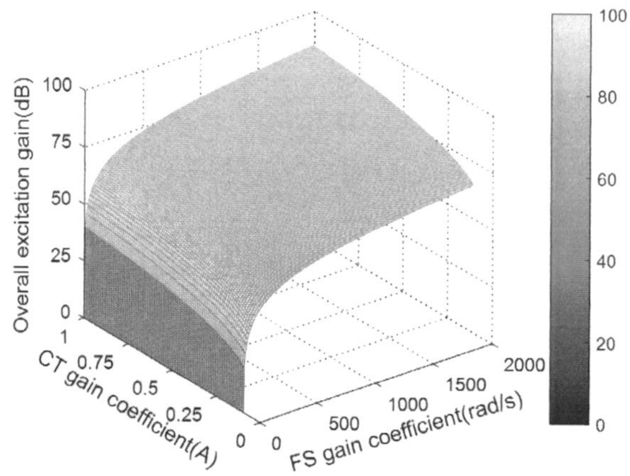

Figure 5. The total excitation gain is displayed when k_m is 0.1

In size-sensitive situations, it is often necessary to limit the aspect ratio of the excitation winding, such as k_m=0.1, or $l_{sol} \cong \widetilde{D}/10$. A two-dimensional surface of the overall excitation gain, as shown in Figure 5, can be obtained. The peak value of the overall excitation gain has been greatly reduced, and the threshold value of the excitation current has been increased. In engineering applications, this type of winding with a very short aspect ratio and high excitation current requires a larger diameter of enameled wire with a small total number of turns, which is suitable for positioning in a shielded medium environment.

5. Conclusions

In this paper, a fresh pulse excitation circuit model is proposed. Taking the traditional resonant mode as a reference, theoretical calculation of the gain of the original excitation model compared to the traditional model is performed in swept frequency mode. Combined with engineering applications, three-dimensional numerical surfaces are generated for the gain results, and the following conclusions are obtained:

(1) As the excitation current increases and the frequency range used outside of the fundamental frequency increases, the new excitation circuit model will have better amplitude-frequency characteristics compared to traditional models, with prominent advantages in applications such as multi-frequency protocols and perceptual positioning;

(2) The different excitation circuit model proposed in this paper has a simple structure, and because resonant capacitors are omitted in the design, the software communication protocol is not limited to the fundamental frequency, with stronger versatility and software and hardware compatibility.

(3) To guide the production of an excitation winding engineering prototype and combine technology with calculation, the concept of length-to-diameter ratio is adopted in this paper. During the numerical analysis process, the impact of unusual length-to-diameter ratios of the excitation winding on the final gain is analyzed in turn, and the issues to be noted in their respective applications are pointed out.

(4) In the ensuing research, the theoretical model of the excitation winding can be deeply explored, and necessary simplification can be made according to the engineering application and process characteristics. Finally, the optimal design scheme of the pulse excitation winding can be obtained, and the theory and process deviation can be explored.

References

[1] I. F. Akyildiz, P. Wang, et al., "Realizing underwater communication through magnetic induction" IEEE Communications Magazine, Vol. 53, No.11, pp.42-48, November 2015.doi:10.1109/mcom.2015.7321970.;
[2] J.H. Jiang et al., "Capacity and Bandwidth Analysis of Symmetric Two-Coil Resonant Magnetic Communication Using Frequency Divarication," IEEE Antennas and Wireless Propagation Letters, Vol. 14, 2015. DOI:10.1109/LAWP.2014.2365556
[3] Z. Sun, I. F. Akyildiz, "Magnetic Induction Communications for Wireless Underground Sensor Networks," IEEE Trans. Antenna and Propagation, Vol.58, No.7, 2010, pp.2426-2435. DOI: 10.1109/TAP.2010.2048858
[4] S. C. Lin et al., "Distributed Cross-Layer Protocol Design for Magnetic Induction Communication in Wireless Underground Sensor Networks," IEEE Trans. Wireless Commun., Vol. 14, No. 7, 2015, pp. 4006-19. DOI:10.1109/TWC.2015.2415812.
[5] A. I. Sasaki et al., "Magnetic-Field-Based Position Sensing Using Machine Learning," IEEE Sensor Journal, Vol. 20, No. 13, July 2020, pp. 7292-7302. DOI:10.1109/JSEN.2020.2979071.
[6] S. Kazeminasab, M. K. Banks, "Localization and Navigation Method for In-Pipe Robot in Water Distribution System," IEEE Access, Vol. 9, 2021, pp. 117496-117511. DOI:10.1109/ACCESS.2021.3106880.
[7] N. Saeed, A. Celik, et al, "Underwater optical wireless communications, networking, and localization: A survey," Ad Hoc Networks, Vol. 94, No. 27, June 2019, pp. 1-35. DOI: 10.1016/j.adhoc.2019.101935.
[8] R. Worley et al., "Robot Localization in a Pipe Network using a Particle Filter with Error Detection and Recovery in a Hybrid Metric-Topological Space," Research Gate Conference Paper. Sep. 2021, pp. 1-9, DOI:10.1109/MFI52462.2021.9591168.
[9] R. A. Khalil et al., "Optimal Relay Placement in Magnetic Induction Based Internet of Underwater Things," IEEE Sensors Journal, Jan 2020, pp.1-8. DOI: 10.1109/JSEN.2020.3012782.
[10] S. Kazeminasab et al., "Localization Techniques for In-pipe Robots in Water Distribution Systems," 2020 8th International Conference on Control, Mechatronics and Automation (ICCMA), pp. 6-11. DOI: 10.1109/iccma51325.2020.9301560.
[11] Z. Li et al., "Underwater cooperative MIMO communications using hybrid acoustic and magnetic induction technique," Computer Networks, Vol.173, 2020, pp.1-14. DOI: 10.1016/j.comnet.2020.107191.
[12] D. Zhao et al., "High Precision In-Pipe Robot Localization with Reciprocal Sensor Fusion-19516," arXiv-CS-Robotics, Feb 2020, pp. 1-12. DOI: arxiv-2020.12408.
[13] G. Liu, "Data Collection in MI-Assisted Wireless Powered Underground Sensor Networks: Directions, Recent Advances, and Challenges," IEEE Communications Magazine, April 2021, pp.132-138. DOI:10.1109/mcom.001.2000921.
[14] Y. Li, S. Wang, C. Jin, Y. Zhang et al., "A Survey of Underwater Magnetic Induction Communications: Fundamental Issues, Recent Advances, and Challenges," IEEE Communications Surveys & Tutorials, Jan 2019, pp. 1-22. DOI:10.1109/COMST.2019.2897610.
[15] K. Miyasaka et al., "Long-mover: Flexible Tube In-pipe Inspection Robot for Long Distance and Complex Piping," 2018 IEEE/ASME International Conference on Advanced Intelligent Mechatronics (AIM), July 2018, pp.1075-1080. DOI: 10.1109/aim.2018.8452690.
[16] D. Wei, L, Yan, X. Li, Y Sun, D Yuan et al., "Exploiting Magnetic Field Analysis to Characterize MI Wireless Communications in Subsea Environments," 2018 International Conference on Computing, Networking and Communications, Wireless Networks, pp. 805-809. DOI: 10.1109/iccnc.2018.8390295.

[17] H. Guo et al., "Practical Design and Implementation of Metamaterial-Enhanced Magnetic Induction Communication," IEEE Access, May 2017, pp.17213-17229. DOI: 10.1109/access.2017.2719406.
[18] A. R. Silva, et al., "Design and Implementation of Low-Power and Mid-Range Magnetic-Induction-Based Wireless Underground Sensor Networks," IEEE Transactions on Instrumentation and Measurement, Vol.65, 2016, pp. 821-835. DOI: 10.1109/tim.2015.2498918.
[19] H. Guo, Z. Sun, J. Sun, N. M. Litchinister, "M2I: Channel Modeling for Metamaterial-Enhanced Magnetic Induction Communications," IEEE Transactions on Antennas and Propagation, Vol.63, 2015, pp. 5072-5087. DOI: 10.1109/TAP.2015.2480095.

Design and testing of automatic downpipe equipment for deep grain piles

Lei Ming[1a], Jiahong Zheng[2, b*], Yue Shi[3] and Jiahou Dong[4]

[1] College of Mechanical and Electrical Engineering, Shaanxi University of Science and Technology, Xi'an City, Shaanxi Province, 710016, China

[2]College of Mechanical and Electrical Engineering, Shaanxi University of Science and Technology, Xi'an City, Shaanxi Province, 710106, China

[3] College of Mechanical and Electrical Engineering, Shaanxi University of Science and Technology, Xi'an City, Shaanxi Province, 710106, China

[4] College of Mechanical and Electrical Engineering, Shaanxi University of Science and Technology, Xi'an City, Shaanxi Province, 710106, China

[a]2388754559@qq.com, [b*]200512091@sust.edu.cn

Abstract. China is a large grain storage country, and at this stage, the effective storage capacity of the country has exceeded 900 million tonnes. Grain is subject to localized high temperatures, pests, and other abnormalities during storage. Therefore, regular quality checks need to be carried out on the grain, and cuttings are the first step in carrying out grain quality checks. The current domestic cutting equipment suffers from low cutting efficiency and high manual labor intensity, so the need to develop new cutting equipment is imminent. In response to the current problems with domestic cuttings equipment, this work has developed a deep grain pile automatic down tube equipment, which is driven by compressed air and is capable of completing cuttings within a ten-meter depth range in grain silos. Compared with traditional cutting equipment, this equipment greatly improves the efficiency of cutting samples, reduces manual labor intensity, provides new ideas for the development of automation of cutting equipment, and has a broad market.

1. Introduction

China is a large grain storage country, and at this stage, grain in storage is mainly stored in vertical silos, shallow round silos, and cottage silos [1]. In the process of grain storage, grain in the silos needs to be sampled regularly in accordance with national standards to ensure the quality and safety of the grain. The current mainstream method is to press the cuttings into the required position in the grain layer by hand with the aid of a pneumatic cutter [2,3]. However, this method is not only time-consuming and laborious, but for deep cuttings of more than five meters, it is difficult to insert them into the required position by hand alone [4,5]. In response to this phenomenon, scholars at home and abroad have researched and developed a variety of grain sampling equipment that can complete the work of cutting samples in grain silos [6-8], but there are problems, such as the large size of the equipment and more complex operation [9-11].

In response to the drawbacks of the traditional cutting method, a portable, movable automatic tube lowering equipment for deep grain piles has been developed, which is capable of taking deep cuttings within a depth range of ten meters. It will automate the process of taking cuttings in silos, reduce manual labor intensity, improve the efficiency of the cuttings, and provide effective technical support for the grain-cutting process.

2. Structural design of automatic downpipe equipment for deep grain piles

2.1. General structural design

The equipment mainly consists of four parts: the lifting mechanism, the rotating mechanism, the base, and the cutting tube. The overall structure of the equipment is shown in Figure 1. The base provides support for the whole machine and enables it to be fixed to the grain surface using the base. The lifting mechanism enables up-and-down reciprocating movements to press the cuttings into and out of the grain layer. The rotating mechanism enables the cuttings to move vertically and at the same time to make a rotating movement, which loosens the "lumpy areas" in the grain layer and thus makes it easier to move the cuttings through the grain layer. Due to the flammability of the grain in the silo, to avoid fires caused by electrical sparks from the motor and the electrical circuit, the equipment uses compressed air as the power source. It is controlled by manual air valves on the lifting and rotating mechanism.

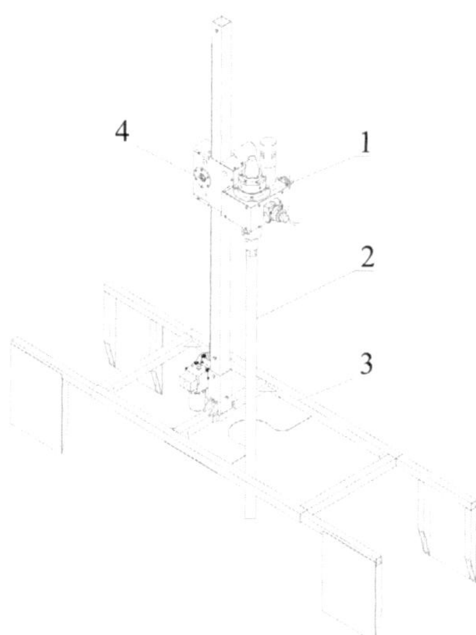

1 Rotating mechanism 2 Sampling tubes 3 bases 4 Lifting mechanism

Figure 1. General structure of the equipment

2.2. Structural design of the base

The base system consists of a support frame, removable baffles (four pieces), and a filter. The support frame is made of welded high-strength aluminum alloy, which reduces the overall weight while ensuring structural strength. Its main function is to provide support for the entire equipment. The removable baffle can be fitted separately into the support frame, and the other end can be inserted into the grain pile to further secure the support frame to the grain surface. The removable baffle is also made of aluminum alloy and is lightweight and easy to carry. The base structure is shown in Figure 2.

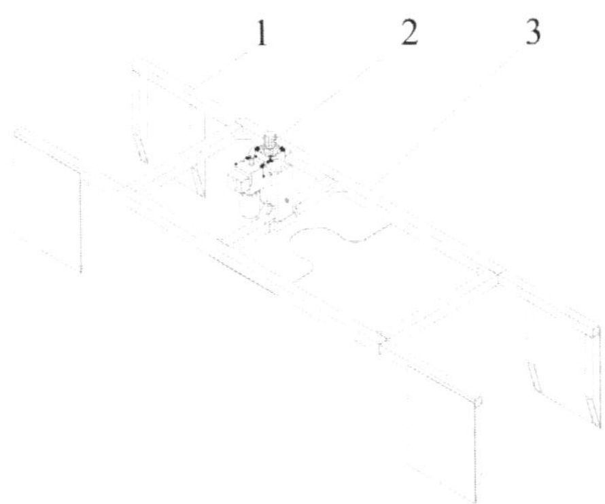

1 Removable stopper (four pieces) 2 Filters 3 Support bracket
Figure 2. Structure of the base

The four removable baffles not only secure the base to the grain surface but also perfectly counteract the torque exerted on the unit during the rotation of the cuttings, thus allowing for a quick lowering of the tubes. The filter is fixed to the support frame and filters out impurities and water from the compressed air, protecting the pneumatic motor on the unit from moisture and impurities in the compressed air.

2.3. Structural design of the lifting mechanism

The lifting mechanism enables up-and-down reciprocating movements to press and pull the cuttings into and out of the grain layer. The lifting mechanism consists of a column, a rack and pinion, a lifting table, and a pneumatic motor. The column is fixed to the base, the rack is welded to the column, and the lifting table is driven up and down by the gears inside the table, which are driven by the pneumatic motor. The structure of the lifting mechanism is shown in Figure 3. The material of the parts contained in the lifting mechanism is aluminum alloy. The column is the key component of the whole equipment, which needs to bear the axial force and bending moment of the cuttings on the equipment. The unique rectangular structure of the column can effectively carry the reaction force of the cuttings on the equipment during the process of lowering the tube to ensure the normal operation of the equipment. The pneumatic motor moves the lift table up and down by driving the gears forward and reverse. The pneumatic motor uses filtered compressed air as the driving source, which is environmentally friendly and pollution-free. In addition, due to the compressibility of the air, the pneumatic motor can automatically stop running when the equipment encounters a fault, effectively protecting the equipment. Eight guide wheels can restrain the movement of the lift.

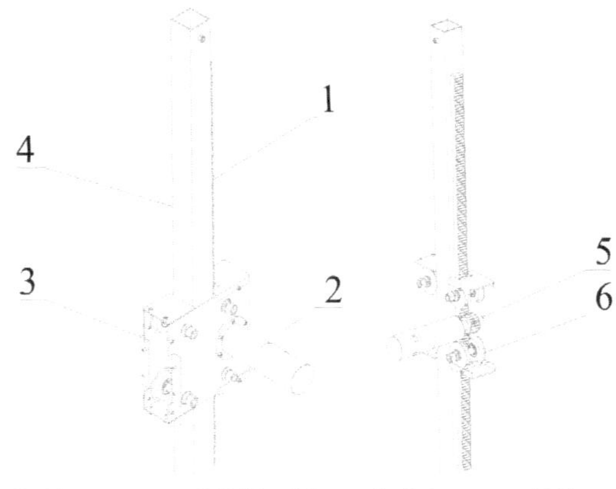

1 Rack and pinion 2 Air motors 3 Lift tables 4 Columns 5 Gears 6 Guiding wheels
Figure 3. Schematic diagram of the structure of the lifting mechanism

2.4. Structural design of the rotating mechanism

The rotating mechanism rotates the cuttings while the tube is lowered, thus loosening the clumped grain and reducing the axial force when lowering the tube. The rotating mechanism is fixedly connected to the lifting mechanism by screws. The rotating mechanism consists mainly of the grain suction port, pneumatic motor, manual air valve, adapter number one and two, pinion gear, large gear, and rotating spindle. The above components, except for the gear, are machined from aluminum alloy and are of reliable construction. The schematic diagram of the structure is shown in Figure 4.

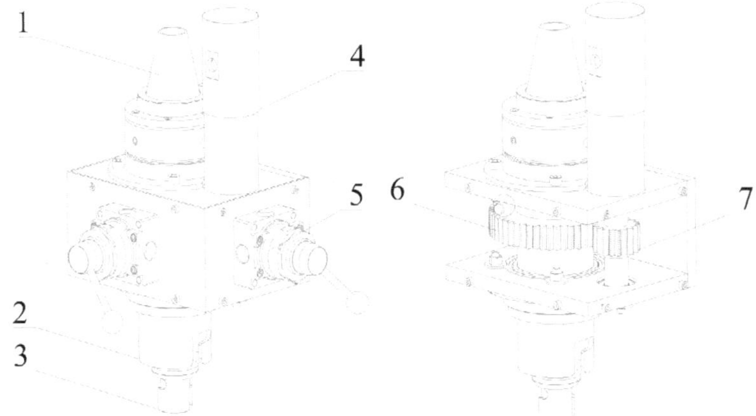

1 Grain suction port 2 Adaptor No. 1 3 Adaptor No. 2 4 Air motors
5 Manual air valve 6 Large gears 7 Pinion

Figure 4. Schematic diagram of the structure of the rotating mechanism

The rotating mechanism contains two sizes of adapters, which can be used to connect different sizes of cutting tubes. It can be adapted to a variety of cutting situations. The grain inlet is designed as a conical structure, which can be quickly connected to the grain suction hose of the cutter. The rotating spindle is below the grain inlet. The large gear is fixedly mounted on the rotating spindle, and the large gear meshes with the small gear. The pneumatic motor powers the pinion gear, which in turn rotates the large gear, and the adapter, Adaptor 1 and Adaptor 2, ultimately give the cuttings a rotational movement. In addition, there are two manual valves on the outside of the rotating mechanism, which control the up and down of the lifting mechanism and the forward and reverse rotation of the rotating mechanism, respectively, enabling precise control of the machine's movements. This improves the efficiency of the cuttings and mechanizes and automates the process.

3. Simulation analysis

3.1. Static simulation

In order to investigate the force situation of the key parts of the deep grain pile automatic down tube equipment and to ensure that the parts will not break during the working process, a static simulation analysis was carried out on the No. 2 adapter in the rotating mechanism of the equipment. The results are shown in Figure 5. The second adapter was made of 7075 aluminum alloy, to which a force of 2000 N was applied to simulate the reaction force during the cuttings. The results show that the maximum equivalent force on adapter No. 2 is 18.199 MPa, which is much less than the yield limit of 430 MPa for 7075 aluminum alloy, indicating that the structure and material selection of adapter No. 2 are reasonable and will not cause failure during operation.

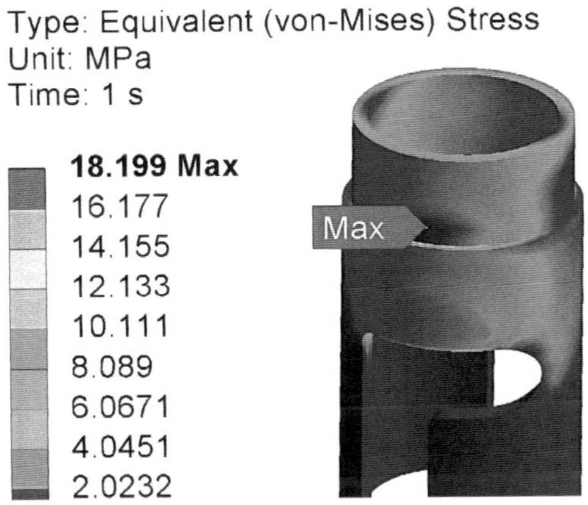

Figure 5. Stress results for adaptor No. 2

3.2. Dynamics simulation

In order to investigate the motion characteristics of the internal gear set of the rotating mechanism of the equipment during operation, the dynamics of the gear set in the rotating mechanism were simulated. The simulation results are shown in Figure 6. The pinion gear material is 40 Cr, and the large gear material is 45 steel. As the pinion gear is the active gear, a torque of size 20 N·m is added to the pinion, and a rotational speed in the opposite direction of the torque to the large gear. From the results, it can be seen that the maximum stress occurs on the contact surface of the gear, with a maximum stress of 250.88 Mpa. This s less than the yield limit of the gear material, satisfying the requirement.

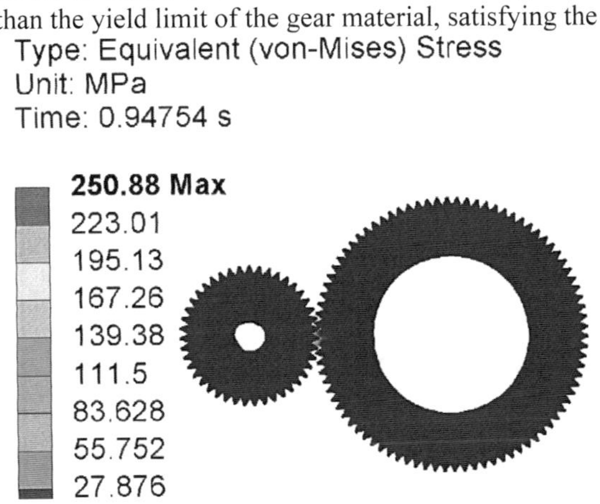

Figure 6. Stress results for the gear set

The variation of the maximum stress on the gear set with time over a time range of 1 s is shown in Figure 7. The results show that the maximum stress changes continuously with the meshing and separation of the gear teeth. The trend of the maximum stress shows an obvious periodicity, which is consistent with the actual motion characteristics of the gear, indicating that the simulation results have a high reference value.

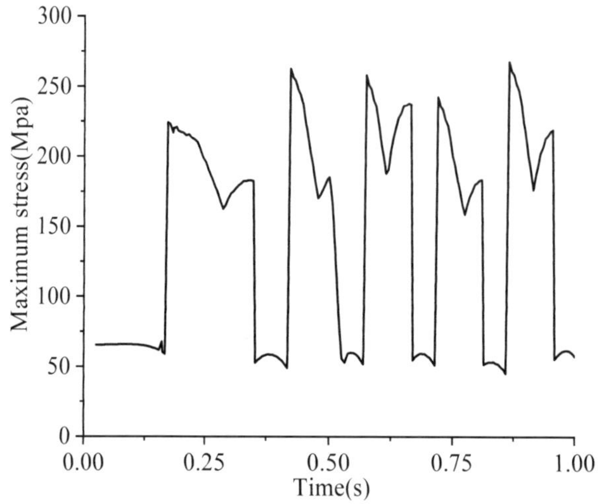

Figure 7. Graph of maximum stress variation

The vertical movement of the machine's lifting mechanism is provided by its internal rack and pinion mechanism, so the movement characteristics of its internal gears have a crucial influence on the stability of the machine's operation. The dynamics of the gear were simulated with the gear speed set at 10 r/min, and a force of 2000 N was set to simulate the reaction force on the device. The results of the torque applied to the gear are shown in Figure 8. The results show that the minimum torque applied to the gear in 10 s is 19909.97 N·mm and the maximum torque is 19910.48 N·mm, with a sinusoidal variation of torque. The reason for this is that the distance between the central axis of the gear and the contact area of the tooth surface constantly varies as the teeth mesh and separate, resulting in variations in torque. The difference between the maximum torque and the minimum torque is 0.51 N·mm, which is small and does not cause large vibrations in the vertical movement of the lifting mechanism and meets the requirements.

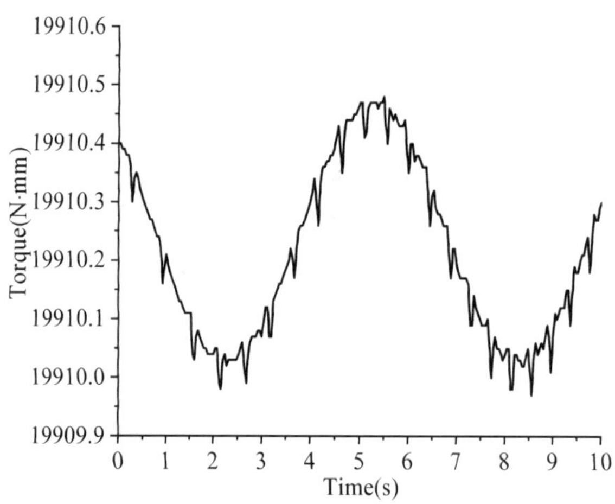

Figure 8. Diagram of the torque applied to the gear

The variation in the value of the force applied by the gear to the rack in the vertical direction as a function of time is shown in Figure 9. The results show that the maximum force applied to the rack is 884.900857 N. The difference between the maximum and minimum force is 0.004138 N, which is a small variation and does not cause fatigue cracks on the tooth surface of the rack due to the variation in force.

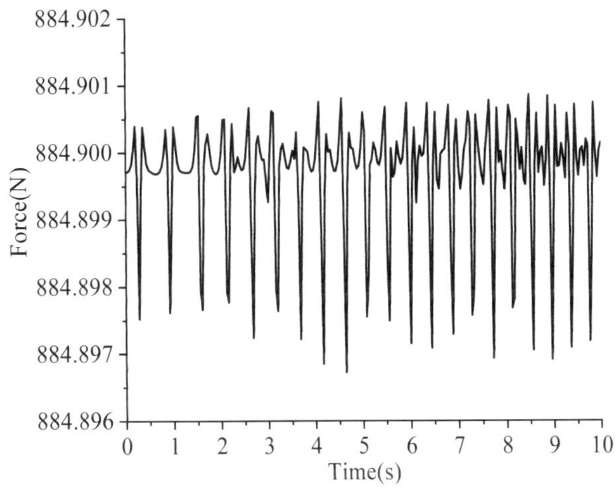

Figure 9. Value of the vertical force on the rack

4. Testing of equipment

In order to investigate the operation and efficiency of the automatic deep pile-down tube equipment in real-life conditions, three randomly selected points in the bins containing rice and wheat were tested, and the reaction forces on the equipment were measured, as shown in Figure 10. A connection device between the deep pile automatic down tube device and the cutting tube, containing two tensile pressure sensors, allows the reaction force on the device to be transmitted via a digital-to-analog converter to the software of the host computer to obtain data during the cutting process.

1 Connecting device
2 Stainless steel cuttings
3 Automatic downpipe equipment for deep grain piles
4 Digital to analogue converters
5 Switching power supplies
6 Upper computer software

a) General structural diagram of the test rig

1 Upper connection plate
2 Tensile pressure sensors
3 lower connection plates

b) Detail drawing of the connection unit

Figure 10. Diagram of the test set-up

The resulting reaction force on the equipment when pressing the cuttings into the grain layer is obtained as shown in Figure 11. The results show that with the use of automatic tube lowering equipment for deep grain piles to press cuttings into the rice layer as well as the wheat layer, the reaction force on the equipment increases with increasing grain depth. Moreover, there is a good linear relationship between the reaction force and the grain depth. The maximum reaction force on the equipment was 594.6 N in the rice layer and 1295.3 N in the wheat layer, which was less than the preset force values in the simulation results, indicating that the above No. 2 adaptor would not have failed during the cuttings.

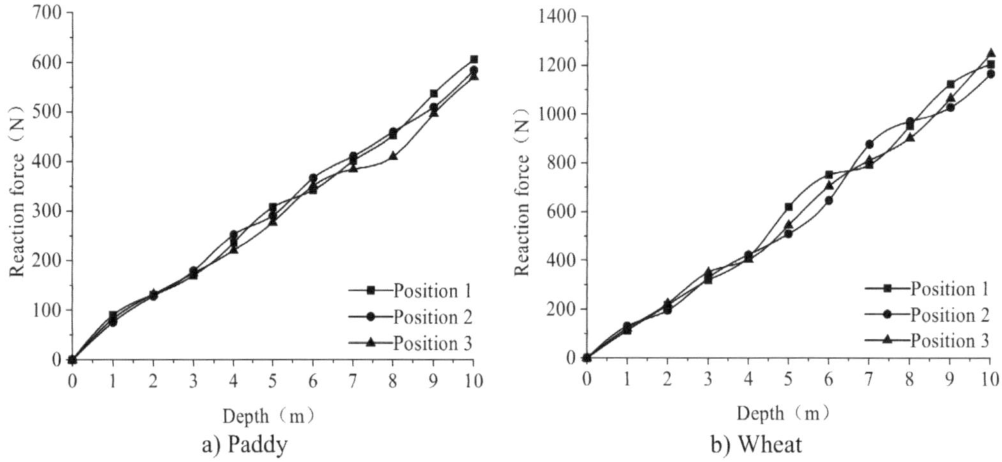

Figure 11. Reaction forces on the equipment when pressing the cuttings into the grain layer

The time taken to lower and pull out the tubes of the deep grain pile automatic tube lowering equipment and the traditional cutting equipment is shown in Table 1. The results show that the time required to complete the cutting process in the rice and wheat silos is less than that of the traditional cutting equipment, thus automating and mechanizing the cutting process in the silos and improving the efficiency of the cutting process.

Table 1. Pipe laying time

Equipment name	Deep grain pile automatic downpipe equipment		Traditional sampling equipment	
Types of cereals	Paddy	Wheat	Paddy	Wheat
Position 1	9.6 min	9.3 min	10.5 min	12.7 min
Position 2	9.4 min	9.6 min	10.8 min	11.9 min
Position 3	9.7 min	9.5 min	11.6 min	12.5 min
Average	9.6 min	9.5 min	11 min	12.4 min

5. Conclusion

Grain storage is an important step in ensuring international livelihoods. Cuttings are an important means of obtaining information on grain quality in storage. To address the problems of low efficiency and labor intensity of traditional sampling methods, in this paper, a deep grain pile automatic down tube equipment is developed. The key components are simulated statically and dynamically. The simulation results show that the structure of the key components meets the requirements. In the actual test process, this equipment, compared to traditional cutting equipment, can quickly complete the deep cuttings in the grain bin. It greatly improves the efficiency of the cuttings. Moreover, it makes the cutting process mechanized and automated, providing a strong guarantee for the grain-cutting process in the bin, with a broad market prospect.

References

[1] Sun, D, Y. (2011) Analysis of the importance of cuttings in inspection work.J. Agricultural technology and equipment,16： 51-52. https://doi.org/10.3969/j.issn.1673-887X.2011.08.029.

[2] Cheng, J. (2012) Key technical aspects of seed cuttings.J. Seed Technology,30(08):33-34. https://doi.org/10.3969/j.issn.1005-2690.2012.08.022.

[3] Cheng, S, S., Yi, D, J., Wang, W, Y. (2016) Implementation and application of intelligent cuttings quality inspection system for grain depots.J. Grain and oil storage technology newsletter, 32(06):23-26. https://doi.org/10.3969/j.issn.1674-1943.2016.06.008.

[4] Bo, J, W. (2019) Use and analysis of bulk grain sampling equipment at quayside.J. Food and Food Industry, 26(03): 48-50. https://doi.org/10.3969/j.issn.1672-5026.2019.03.015.

[5] Huang, Y. (2018) Study of the differences in the results of grain impurities and imperfection determination by different cutters.J. Modern food, 19: 88-90+94. https://doi.org/10.16736/j.cnki.cn41-1434/ts.2018.19.028.

[6] Ying, S, J., Ying, Y. (2013) Development of an intelligent electric suction cutter for grain.J. Grain, oil and food technology, 21(06): 21(06): 120-122. https://doi.org/10.3969/j.issn.1007-7561.2013.06.032.

[7] Yang, J, J., Zhang, H, S., Liu, Y, H. (2021) Design and application of deep grain sampler in large granary.J. Grain and feed industry, 01:13-16. https://doi.org/10.7633/j.issn.1003-6202.2021.01.005.

[8] David., Fuchs. (2018) Port-a-Probe Vacuum Grain Sampler: KC Supply's GVS portable vacuum pump makes monitoring stored grain quality easier.J. Grain Journal, 46(2): 70-71. https://www.nstl.gov.cn/paper_detail.html?id=9b8d3ddb85b62166f5d9862c4ab1ee33.

[9] Xing, Y. (2011) Current situation and research direction of domestic grain sampling equipment.J. Grain storage, 40(03):53-56. https://doi.org/10.3969/j.issn.1000-6958.2011.03.012.

[10] Wang, J., Li, H, P. (2021) Effect of optimizing sampler equipment on drying imperfect corn grains in shallow round bin.J. Modern food, 20: 37-39. https://doi.org/10.16736/j.cnki.cn41-1434/ts.2021.20.010.

[11] Li, J, Y., Li, J, P., Cheng, P, F. (2020) Study on the key factors influencing the sampling process on the imperfect grain of dried corn in Northeast China.J. Modern food, 10: 7-10. https://doi.org/10.16736/j.cnki.cn41-1434/ts.2020.10.003.

Journal of Physics: Conference Series **2561** (2023) 012012 doi:10.1088/1742-6596/2561/1/012012

Design and Simulation Analysis of the Patient Transfer Machine

Shaoxin WEN [1,2], Xinzui Wang [1,*], Fucheng Cao [1], Hui Fu [2]

[1]Ji Hua Laboratory, Foshan 528200, China;

[2]School of Mechanical and Electrical Engineering, Guangdong University of Technology, Guangzhou 510006, China

wangxz@jihualab.com

Abstract. In response to the low level of automation of the tools used for patient transfer in hospitals and the resulting shortage of nursing personnel, this paper studies the characteristics of existing transfer products and related structural features. A transfer machine based on static transfer is proposed, which can achieve patient transfer between a hospital bed and a stretcher truck. In this paper, the overall structure of the transfer machine has been designed in detail, and the structural design and analysis of important transmission components have been carried out. The finite element strength analysis of the main bearing components has been carried out. Finally, the selection of driving components based on their functions has been carried out. This study provides theoretical and technical support for the development of subsequent transfer machines.

1. Introduction

As China's aging trend continues to deepen, the number of people with disabilities caused by factors such as age and disease is also increasing, and the issue of how to transport such groups is becoming more prominent [1,2]. Currently, the traditional method of manually carrying patients to stretchers is still used for patient transfer in hospitals, which brings high labor intensity to nursing workers and is prone to secondary injury to patients [3]. Fortunately, various transfer nursing robots have been developed in these years, and a series of studies have been carried out for the analysis of these robots, especially some of the devices that have been clinically proven to be effective[4,5]. As shown in Figure 1(a), the hanging bag type transfer machine [6], after adjusting the patient's posture during the transfer process, accurately wraps the hanging bag around the patient and follows the movement of the electric transfer machine to achieve the purpose of patient transfer. However, this type of transfer machine needs to flip the patient, which can easily cause secondary injury to the patient. Japan has a relatively large number of transportation products and is relatively mature. In 2009, a transportation robot named RIBA [7] was developed. As shown in Figure 1(b), the robot has two arms, which simulate the insertion of a human arm under the patient's body to move. However, this method is not suitable for patients with spinal injuries or fractures. There are also relatively mature products in China, as shown in Figure 1(c). The LW-TB-1 transfer machine was developed by Dalian Longwei Medical Co., Ltd [8]. The device can accurately and automatically transfer patients, reducing the labor intensity of nursing staff. However, due to its large volume and high working environment requirements, the transfer machine cannot be separated from the walking equipment.

In response to the above issues, this article proposes a single-person operable transfer machine based on modular thinking. This device is separated from a stretcher truck, which is easy to operate, and can be transferred without changing the patient's posture, avoiding secondary injury to the patient.

(a) (b) (c)
Figure 1. Transfer product

2. Working principle and structural design of transfer machine

2.1. Working principle of the transfer machine

The transfer device needs to complete the transfer of patients between the hospital bed and the stretcher truck without changing the patient's posture. To achieve this goal, the transfer device is designed into two parts, namely, a transfer module and a walking module. Both parts realize their functions through a belt transmission. The belt movement of the transfer module can drive the movement of the human body, and the belt movement of the walking module can make the shipping device itself produce the translation. The specific workflow of the shipping device is shown in Figure 2:

(1) The transfer module does not move, while the walking module moves through a belt to pan from the stretcher truck to approach the patient, as shown in Figure 2(a).

(2) The transfer module and the walking module belt move simultaneously, placing the patient on the transfer machine through belt friction, and then the walking module works independently to translate to the stretcher truck, as shown in Figure 2(b).

(3) After the transfer machine transfers the patient to the stretcher truck, the transfer belt and the walking belt work simultaneously to make the transfer machine withdraw between the stretcher truck and the patient, and the transfer machine moves towards the hospital bed, as shown in Figure 2(c).

(4) As shown in Figure 2(d), the transfer machine returns to the hospital bed, and the entire transfer process from the stretcher truck to the hospital bed is completed, as shown in Figure 2(d).

The above is the process of transferring a patient from a hospital bed to a stretcher truck, while the process of transferring a patient from a stretcher truck to a hospital bed is the opposite.

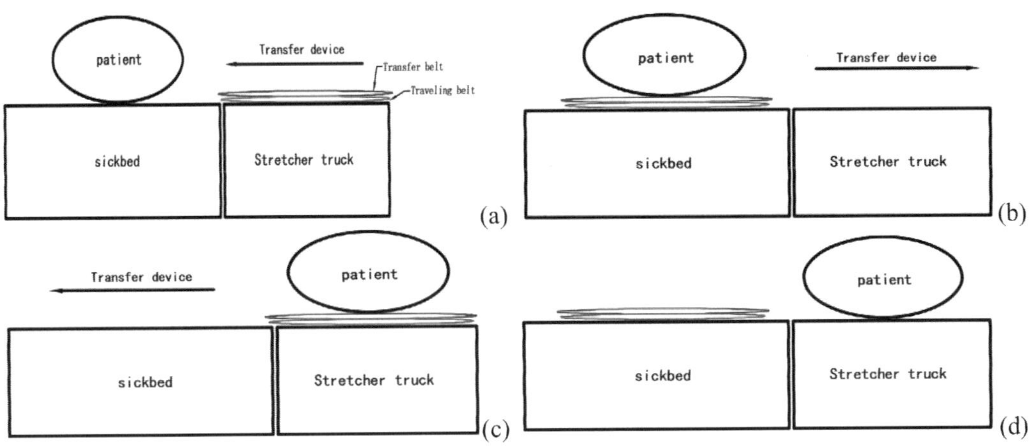

Figure 2. Transfer Flow Chart

2.2. Structural design of transfer machine

In response to the actual needs of the hospital in the process of transferring patients, and by referring to the Chinese adult human body data standard analysis [9], the physiological structure of the human body was analyzed, and a new type of mechanism configuration of the transfer device was designed to solve the problems of the high labor intensity of nursing workers and secondary injury to patients during the process of transferring patients in the hospital.

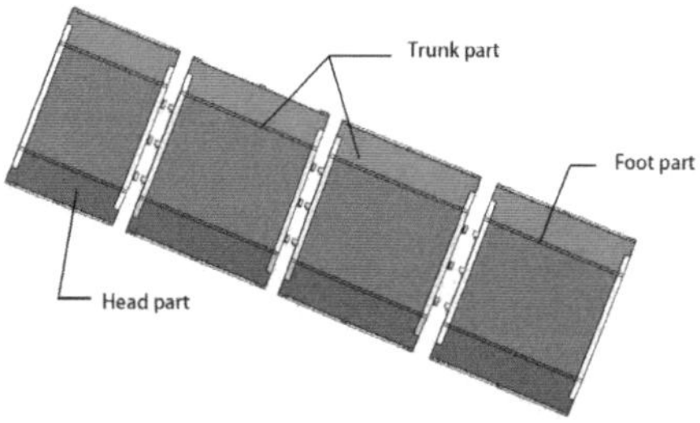

Figure 3. Disassembly drawing of machine

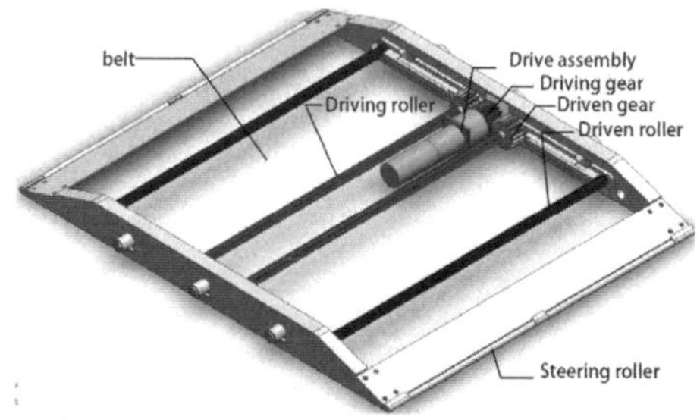

Figure 4. Transfer module structure diagram

As shown in Figure 3, in the width direction, the transfer machine shows that the middle part is planar, and both ends are transitioned with arcs. This design helps to insert better. In addition, to improve the portability of the transfer machine, the overall design of the transfer machine is divided into three parts: the head part, the trunk part, and the foot part, of which there are two trunk parts.

The transfer device's head, torso, and foot parts have approximately the same internal structure except for differences in size. One of the modules can be selected for mechanical structure analysis. In this article, the torso module with a higher load carrying requirements is selected.

The structure of the torso transfer module is shown in Figure 4. The entire transmission structure mainly includes driving gears, from moving gears, driving rollers, motion rollers, flat bands, and synchronous bands. The moving wheel is connected to the active roller.

The belt transmission diagram of the transfer module is shown in Figure 5. The synchronous belt is wound between two pairs of driving rollers and driven rollers to form a synchronous belt winding transmission mechanism. The flat belt is fixed to the two synchronous belts along the transfer direction. During the transfer process, the power components output power, and the driving rollers on both sides are driven to rotate in the same direction, while the synchronous belts on both sides rotate in the same direction. This transmission method effectively reduces the risk of slipping and deviation during the transfer process.

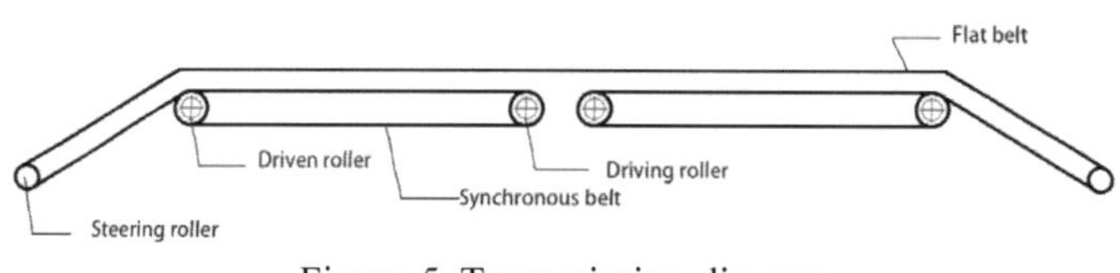

Figure 5. Transmission diagram

The structure of the walking module is shown in Figure 6. The similarity between the driving component of the walking module and the transfer module is that they are both composed of a motor, a coupling, a reducer, and other parts. The difference is that the walking module is directly connected to the driving roller by the motor to transmit power, making the structure more compact and simple. The position of the tensioning roller can be adjusted by tightening the nut to achieve the effect of tensioning the synchronous belt, preventing the traveling belt from jumping teeth during operation, and resulting in poor transmission accuracy.

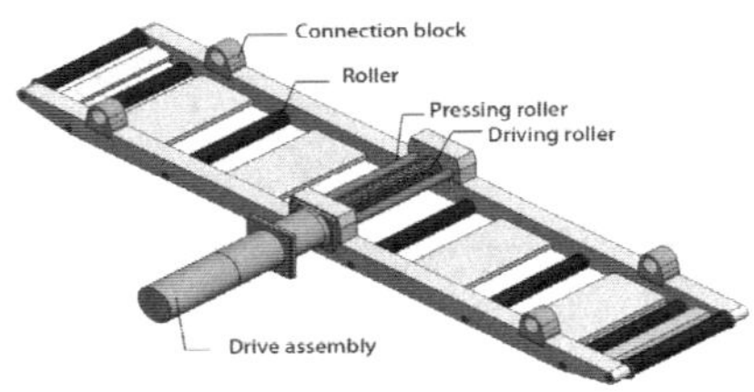

Figure 6. Walking module diagram

3. Selection of drive motor for the transfer device

When the transfer module is working, the maximum patient weight of the device is set to 100 kg. For safety reasons, a 20% margin is given, that is, overcome the gravity m=120 kg. The distance that the patient needs to be lifted is h=0.066 m, the width of the transfer machine is 640 mm, the width of the human body is set to 500 mm, and the moving distance is half of the sum of the two, that is, 570 mm. The flat belt is made of PVC, with a static friction coefficient of 0.4-0.55, and a friction coefficient of 0.6 for the panel (safety) and 0.6 for the human body. In addition, from the perspective of efficiency and comfort, the translation speed v=0.06m/s is taken. The work done to overcome gravity during transport is:

$$W_G = mgh = 120 \times 9.8 \times 0.066 = 77.616J \quad (1)$$

The friction force to be overcome is:

$$F_f = mgf = 120 \times 9.8 \times 0.6 = 705.6N \quad (2)$$

The work done to overcome friction is:

$$W_f = F_f s = 705.6 \times 0.57 = 402.19J \quad (3)$$

The time required for the transfer process is:

$$t = \frac{s}{v} = \frac{0.57}{0.06} = 9.5s \quad (4)$$

The average power is:

$$P_n = \frac{W_G + W_f}{t} \approx 50.5W \quad (5)$$

When a person lies flat, the main forces are on the torso and feet. Therefore, based on the calculation of three motor drives, the power of a single motor is:

$$P = \frac{P_n}{3} = 16.84W \quad (6)$$

The average efficiency of the motor is basically 0.87, which is taken as 0.8 (safe). If the operating efficiency of the reducer is taken as, the power of the motor is:

$$P_t = \frac{P}{\eta_1 \cdot \eta_2} = \frac{16.84}{0.8 \times 0.8} = 26.31W \quad (7)$$

During actual transportation, due to factors such as mechanical friction, heat loss, and current, the actual power required by the motor should be greater than the theoretical power. Considering the limited equipment space, Coriolis hollow cup motor 32SYK6400 was selected, and its motor parameters are shown in Table 1.

Table 1. Motor parameter table

Rated voltage(V)	Rated power(W)	Maximum power (W)	Rated torque (N•mm)	Rated current(A)	Rotate current(A)	Rated speed(r/min)
24	66	169	96.7	3.14	28.3	6580

The outer diameter of the synchronous belt driving roller of the transfer module is 14mm, and the transmission ratio of the driving wheel and the driven wheel of the transfer module is set to 1:1. The torque required to output by the motor is calculated as follows:

$$T = F \cdot \frac{d}{2} \cdot i = \frac{P}{v} \cdot \frac{d}{2} \cdot i = \frac{16.84}{0.06} \times \frac{0.014}{2} \times 1 = 1.96N\ m \tag{8}$$

Due to mechanical losses in the system, a safety factor is added to the output torque, and the added torque is:

$$T_\lambda = T \cdot \lambda = 1.96 \times 1.8 = 3.528N\ m \tag{9}$$

Calculate the required reduction ratio of the reducer based on this torque:

$$i_{DC} = \frac{T_\lambda}{T_n} = \frac{3.528}{0.0967} = 36.48 \tag{10}$$

Where, λ is the safety factor, and T_n is the rated torque of the motor. Checking the relevant manual, the reduction ratio of the transfer module reducer is determined as 42.92.

After selecting the speed reducer, it is also necessary to verify whether the motor speed is within the rated speed when the equipment operates normally at the set speed to avoid damage to the motor. If the linear speed v of the drive roller is 0.06m/s and the outer diameter of the drive roller is 14mm, the motor output speed is:

$$r_d = \frac{60 \cdot v}{\pi \cdot D} \cdot i \cdot u = \frac{60 \times 0.06}{3.14 \times 0.014} \times 42.96 \times 1 = 3518.11 r/\min \le 6580 r/\min \tag{11}$$

During the transfer operation, the rotation speed is within the normal range, so the reducer selection is reasonable.

During the transfer process, when the patient lies flat on the transfer machine, the walking module bears the largest load, so we also need to select the motor for the walking module. As previously known, the safe carrying capacity of the transfer machine is 120kg, the total weight of the transfer machine is 60kg, and its operating speed is v=0.06m/s. During the working process of the walking module, it is necessary to overcome the friction force to do the work. The friction coefficient f of the walking belt and the mattress is set to 0.6. For safety reasons, assuming that the weight of the transfer machine and the patient are all pressed on the bed, the work done by the walking module to overcome the friction force is:

$$W_f = F_f s = fMgs = 0.6 \times 180 \times 9.8 \times 0.57 = 603.3J \tag{12}$$

The total power that needs to be provided during equipment work is:

$$P = F_f v = 1058.4 \times 0.06 = 63.504W \tag{13}$$

The single motor output power is calculated according to three motors:

$$P_{AC} = \frac{P}{3} = \frac{63.504}{3} = 21.168W \tag{14}$$

Other calculation processes are similar to the transfer module and will not be repeated. After checking, it is concluded that the Coriolis hollow cup motor 32SYK6400 is also suitable for the walking module.

4. Design and simulation analysis of key transmission components

4.1. Design Verification and Simulation Analysis of Gears

As an important component in the transfer module that connects the output end of the motor and the drive wheel and transmits power, the gear needs to be designed and checked. Based on the selection of the motor mentioned above and the transmission structure, it can be determined that the transmission power of the driving wheel is the same as the rated power of the motor under the rated voltage operation of the transfer machine, namely: P=66W.

The active gear speed is:

$$n_1 = \frac{n_e}{i_u} = 153.3r/\min \tag{15}$$

The torque of the active gear is:

$$T = 9550 \times \frac{P}{n_1} = 4.11N \cdot m \tag{16}$$

(1)Design according to the fatigue strength of the tooth surface contact

According to the mechanical design manual [10], the gear is manufactured with a precision of Level 8 according to the usage conditions. The material is 45 # steel for surface hardening treatment. The design calculation formula for the contact fatigue strength of the tooth surface is as follows:

$$d_1 \geq A_d \sqrt[3]{\frac{KT_1}{\phi_d \sigma_{HP}^2} \cdot \frac{u+1}{u}} \tag{17}$$

Where d_1 is the diameter of the graduation circle (mm); A_d is the constant contact coefficient of the gear pair; K is the Load factor; T_1 is the transmission torque of the driving gear (N • m); ϕ_d is the tooth width coefficient of the driving gear; σ_{HP} is the allowable contact fatigue stress; u is the meshing transmission ratio of two gears.

Refer to the relevant mechanical design manual and substitute the above parameters into the formula to obtain $d_1 \geq 15.32$ is sufficient, and the diameter d of the indexing circle of the driving wheel is taken into account the actual matching problem d_1=36mm, the number of teeth is taken as z_1=18, the driven wheel is the same as the driving wheel. According to the requirements of gear design standards, the modulus m_1 is:

$$m_1 = \frac{d_1}{z_1} = 2 \tag{18}$$

(2) Stir-rooted fatigue intensity of the school

After designing according to the gear contact fatigue strength, it is also necessary to verify and calculate the bending fatigue strength of the gear, such as:

$$m \geq A_m \sqrt[3]{\frac{KT_1 Y_{Fs}}{\phi_d z_1^2 \sigma_{FP}}} \tag{19}$$

Where A_m is the helix angle coefficient, Y_{Fs} is the tooth profile composite coefficient, and σ_{FP} is the allowable bending fatigue stress.

After checking the relevant manuals to substitute each parameter, we obtain $m \geq 0.863$, and the number m_1 = 2 meets the requirements. The specific parameters of the gear are finally determined in Table 2.

Table 2. Gear parameter table

Name	Tooth	Pressure angle	Division round diameter	Model
Active wheel	18	20°	36	2
Driven wheel	18	20°	36	2

(3)Gear meshing simulation verification

In order to verify whether the design verification of the gear is reasonable, the Adams dynamic simulation software is used to simulate the gear meshing process. After substituting the parameters, the force condition at the meshing point and gear vibration condition of the transfer module at the rated voltage is obtained, as shown in Figure 7 and Figure 8.

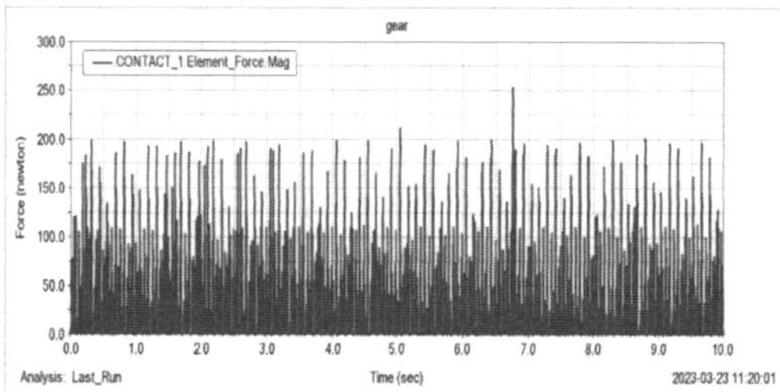

Figure 7. Gear torque variation diagram

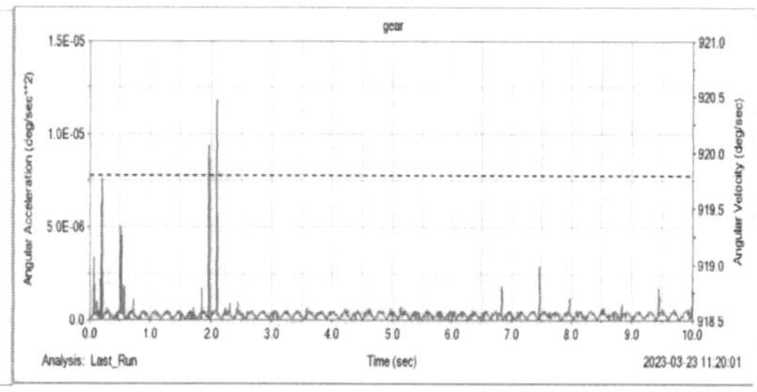

Figure 8. Gear offset variation diagram

As can be seen from the above, its stress and deflection are within an acceptable range; that is, the selected gear meets the requirements, and the driven wheel and driving wheel are the same, so the driven wheel is also safe and reliable.

4.2.Design Verification and Simulation Analysis of Steering Roller

The steering roller in the transfer module is the thinnest and stronger among all transmission rollers, so it is necessary to design and check the school. The schematic diagram of the deflection and bending moment of the turning roller is shown in Figure 9.

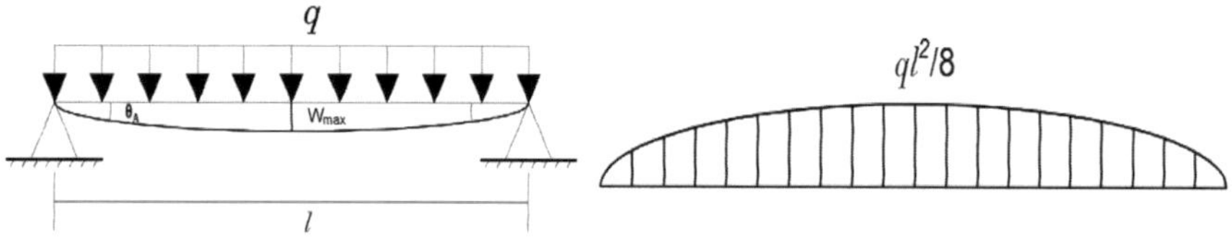

Figure 9. Schematic diagram of bending moment and deflection of the steering roller

From the above bending diagram, the maximum bending moment $M_{\max}$ that turns the roller is:

$$M_{\max} = \frac{ql^2}{8} \tag{20}$$

Among them, the average load of the q is the steering roller; l is the length of the roller, so the deflection equation of the steering roller is:

$$w = -\frac{qx}{24EI}(l^3 - 2lx^2 + x^3) \tag{21}$$

The cross-section corner θ_A and θ_B, and the maximum deflection $W_{\max}$ are obtained, respectively:

$$\theta_A = -\theta_B = -\frac{ql^3}{24EI} \tag{22}$$

$$w_{\max} = -\frac{5ql^4}{384EI} \tag{23}$$

According to the formula, the longer the steering roller, the greater the deflection, and there is a risk of its falling off. Therefore, the steering roller is divided into two sections, each with a diameter of 8mm and a length of 170mm. The material of the steering roller is commonly used 45# steel.

Perform a finite element analysis using Ansys. After applying force and constraint conditions to the steering roller, perform grid division to make its stress converge, and the simulation results obtained are shown in Figure 10. From Figure 10(a), it can be seen that its maximum stress value is 76.75Mpa, which is far less than the yield limit of 45# steel. From Figure 10(b), it can be seen that its maximum deformation is 2.395 × 10^{-2} mm, within the acceptable range. In summary, the designed strength of the steering roller meets the use requirements.

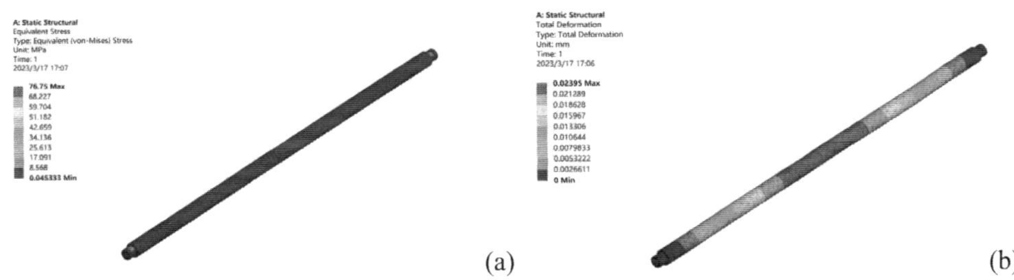

(a) (b)

Figure 10. Simulation analysis diagram of the steering roller

5. Strength verification analysis of bearing parts

5.1.Strength verification analysis of beam

According to the characteristics of the human body lying flat, it can be seen that the stress is basically concentrated in the trunk part, so it is sufficient to check the beam of the trunk part. For safety reasons, it is assumed that all the forces on the human body are concentrated in the torso, which has a total of 4 beams. The patient's weight set in this article is 120kg, so the force on each beam is 294N.

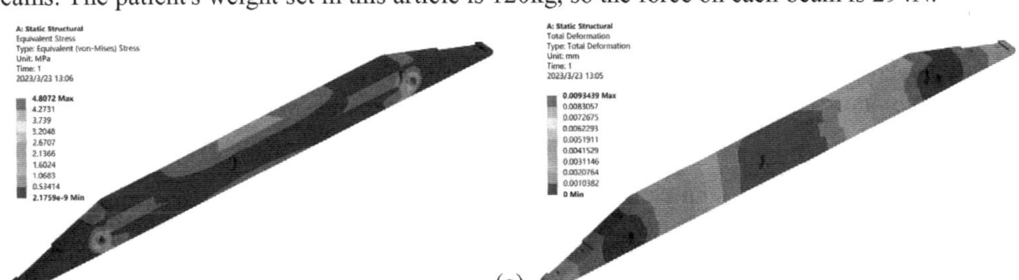

(a) (b)

Figure 11. Strength analysis diagram of the cross beam

In order to reduce the weight of the entire equipment, metal materials with low density should be selected as far as possible, provided that the conditions for use are met. Therefore, the beam is made of aluminum alloy. After performing conditional constraints using the finite element analysis software ANSYS, the convergent simulation results are shown in Figure 11.

From 11(a), it can be seen that the maximum stress on the torso module beam is 4.8072Mpa, which is far less than the yield limit value of aluminum alloy. Figure 11(b) shows that the maximum deformation of the crossbeam occurs in the middle region, with a deformation amount of approximately 0.0093mm, which is within the acceptable range. Therefore, the design of the trunk crossbeam meets the requirements for use, and it is also shown laterally that other crossbeams meet the requirements.

5.2.Strength Verification Analysis of the Connecting Block between Bearing Rod and Frame

In the transfer module, in addition to the beam, the main bearing components also include a bearing rod and a frame connection block. The main function of the load-bearing rod is to carry loads, while the main function of the frame connection block is to connect the transfer module and the walking module.

When the bearing rod contacts the connecting block, certain friction will occur, so the material selected is 440C stainless steel. The frame connection block is made of 304 stainless steel.

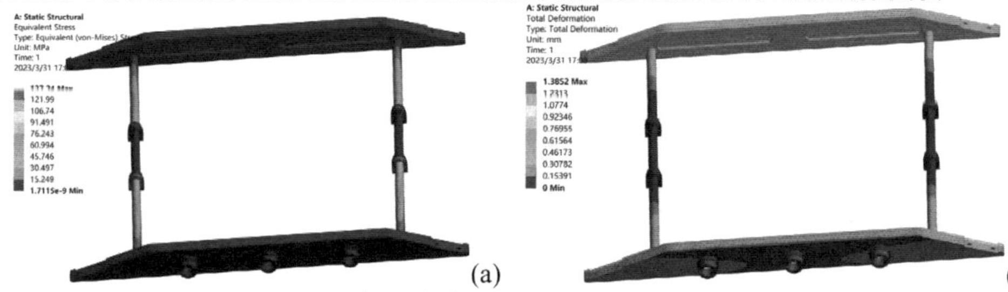

(a) (b)

Figure 12. Strength analysis of connecting block and bearing rod

In order to ensure that finite element analysis can be carried out effectively, it is necessary to omit components unrelated to the analysis. The results of the strength analysis of load-bearing components are shown in Figure 12. From 12(a), it can be seen that the maximum stress on the bearing rod is 137.24Mpa, which is less than the yield strength of 440C stainless steel of 420Mpa. The maximum stress on the frame connection block is about 91.491Mpa, which is less than the yield limit of 304 stainless steel. From 12(b), it can be seen that the deformation of the bearing rod gradually increases from the middle along both ends, with a maximum deformation of about 2.5 mm. The deformation of the frame connection block is smaller and can be ignored. Therefore, both structures meet the design requirements.

5.3.Walking module racks' finite element verification analysis

In the walking module, the walking frame is the main bearing component. In order to reduce equipment weight and economic considerations, the most commonly used 6061 aluminum alloy is selected for the rack material. The total gravity applied to the traveling frame is 1764N, and the area where it contacts the connecting block is the force-bearing area. The force is calculated as eight force-bearing areas, i.e., 220.5N for each area. The simulation results obtained using the software are shown in Figure 13.

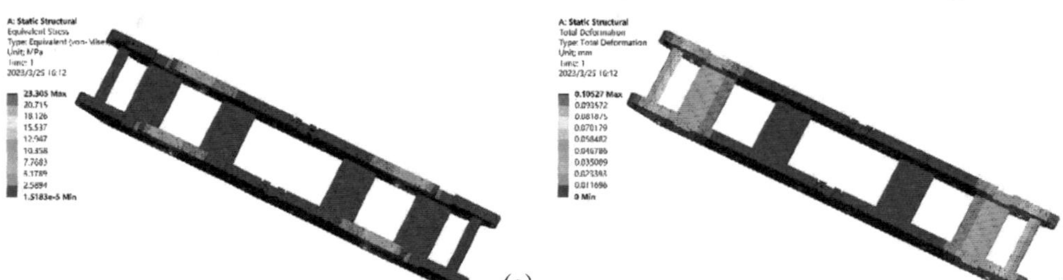

(a) (b)

Figure 13. Strength verification analysis of walking frame

From the stress diagram Figure 13(a), it can be seen that the maximum stress generated by the traveling module frame is 23.576 Mpa, which is lower than the yield limit of 6061 aluminum alloy of 55.2 Mpa. As can be seen from Figure 13(b), the maximum deformation caused by the walking frame is very small, about 0.392mm. Therefore, the design of the trunk walking module frame meets the use requirements.

6. Conclusions

According to the actual needs of hospital transportation, this article proposes a patient transportation machine, expounds its working principle, and carries out structural design. The appropriate motor and reducer for its function are selected, then its key transmission components are designed and simulated. Finally, a finite element strength analysis of the main bearing parts of the equipment is conducted, verifying the feasibility and reliability of the structural design of the transfer machine, which is of great significance for the subsequent development of the transfer machine.

Acknowledgment
Supported by the Jihua Laboratory of China (Grant No. X190341TD190); 2020 Sui Talent Program No. 9(Grant No. 202009030007); Jilin Province and Chinese Academy of Sciences Science and Technology Cooperation High-tech Industrialization Project [2021SYHZ0020].

References

[1] Li, T., Wang, X. (2016) The Connotation and Realization Path of Healthy China. Health Economics Research 33 (1)： 4-910. 10.14055/j.cnki.33-1056/f.20160106.002.
[2] China's National Health Commission. (2021) In 2021, the national residents' health literacy level will reach 25.40%. https://baijiahao.baidu.com/s?id=17350491765 95083373&wfr=spider&for=pc.
[3] He, M., Yao, Y., Lin, C., Zhong, M. (2020) Research progress and development trend of t ransfer nursing robot. Mechanical Design and Manufacturing No.348(02):162-165+169.DOI:10.19356/j.cnki.1001-3997.2020.02.040.
[4] K. Krishnaswamy, S. Moorthy, and T. Oates. (2017) Survey data analysis for repositioning, transferring, and personal care robots in Proc. 10th Int.Conf. Pervasive Technol. Rel. Assistive Environ., Island Rhodes, Greece, Jun. pp. 45–51.
[5] M. Vander Loos and D. Reinkensmeyer, (2008) Rehabilitation and health care robotics, in Handbook of Robot, K. Siciliano, Ed. New York, NY, USA: Springer, ch. 53, pp. 12 23–1249.
[6] Swedish Anjiu Company. (2017) Ceiling Lift Systems from Arjo Huntleigh Enable Residen ts/Patients to be Transferred in a Safe, Comfortable and Dignified Way[EB/OL]. http://www.arjohuntleigh.com/products/patient-transfer-solutions.
[7] Onishi, M., Luo, Z., Odashima, T. (2007) Generation of Human Care Behaviors by Huma n-Interactive Robot RI-MAN[C]//Proceedings of IEEE International Conference on Net working, Robotics and Automation, Roma,3128-3129. http://doi.org/10.1109/ROBOT.20 07.363950.
[8] H.Wang, L. Xie, X. Zhao. (2015) Mechatronics System Design and Experiment Research f or a Novel Patient Transfer Apparatus[C].3rd International Conference on Mechatronic s,Robotics and Automation(ICMRA),77(4):1140-1143.http://doi.org/10.2991/icmra-15.201 5.220.
[9] China Technical Supervision Bureau. (1988) Human Dimensions of Chinese Adults (GB/T 10000-1988). http://c.gb688.cn/bzgk/gb.
[10] Cheng, D. (2016) Gear driving. In: Da Xian, Cheng, Mechanical Design Manual. Chemic al Industry Press, Beijing15-23.

Journal of Physics: Conference Series **2561** (2023) 012013 doi:10.1088/1742-6596/2561/1/012013

The Arrangement and Cross-sectional Shape of Cooling Fins on Train Brake Discs

Hao YU[1,a],Chun TIAN[1], Guozhuang WANG[1]

[1]Institute of Rail Transit, Tongji University, Shanghai, China

[a]2131399@tongji.edu.cn

Abstract-Taking into account physical factors such as the wind effect of the brake disc pump, heat dissipation power, and temperature rise, the optimal cross-sectional shape and layout of the brake disc cooling fins are studied. The flow field distribution and pump-wind heat dissipation power ratio of four different cross-sectional shapes, namely circular, elliptical, square, and rhombic, are compared and analyzed. The flow characteristics, wind effect, and heat dissipation power of the cooling fins under two different layouts, namely crosswise and parallel layouts, are also analyzed. The research results show that the circular cooling fins have the optimal pump-wind loss and heat dissipation power, and the crosswise layout of cooling fins shows the best comprehensive performance. The performance analysis method of cooling fins obtained in this paper provides a reference for the optimization design of brake disc structures.

1. Introduction

The brake disc of a high-speed train is an important braking component, which converts the kinetic energy of the train's movement into heat through the frictional braking action between the brake disc and the brake pad, and dissipates it through the contact with the surrounding airflow to slow down or stop the train. Currently, most high-speed train brake discs use ventilated brake discs. Although the use of ventilated brake discs can effectively increase the heat dissipation power, the complex flow environment under the train causes the exacerbation of the pump wind effect of the ventilated structure, resulting in a significant energy loss. Therefore, while improving the heat dissipation capacity of the brake disc, it is urgent to study how to reduce the air-resistance energy consumption of the brake disc.

To address this issue, scholars at home and abroad have conducted extensive research. Zuo et al. [1] established a model of the train body, brake disc, bogie, and the surrounding flow field, and used the fluid-solid coupling finite element simulation method to calculate the pump wind power consumption of the high-speed train brake disc. Their research analyzed the brake disc pump wind effect but did not discuss the heat dissipation characteristics of the brake disc. Based on this, Zuo et al. [2] used ANSYS to establish a finite element model of the brake disc and studied the influence of brake disc structural parameters such as heat dissipation rib shape and diameter on the temperature field based on the moving heat source method. Ma et al. [3] used the finite element software MARC to simulate the temperature rise process of the brake disc during the train braking process and analyzed the interface stress changes caused by temperature. Zhang [4] used a four-factor orthogonal experimental design to calculate the pump wind power and heat dissipation power of the elliptical heat dissipation rib for optimal design. Suo et al. [5] determined the thermal-elastic-plastic constitutive relationship of the train brake disc through experiments and simulated the stress-strain response relationship and temperature field distribution of the brake disc under various braking conditions.Kravchenko et al. [6]calculated the airflow resistance of train brake discs through a test rig and computer simulation, and proposed that the key parameter for brake disc cooling is the airflow in the cooling channel. Luo et al.[7] calculated the distribution of the convective heat transfer coefficient on the surface of the brake disc through the conjugate heat transfer method, and proposed that the heat dissipation characteristics of the brake disc can be improved by reducing the thickness of the heat transfer wall of the brake disc.Wang et al.[8] used a sequential coupling method to simulate the temperature field of the brake disc under heat dissipation conditions, and loaded complex hexagonal thermal loads. Finally, it was verified that the calculation error did not exceed 5%. You et al.[9] conducted a numerical study on the effect of wheel rotation on underflow flow and aerodynamics of high-speed trains using the Reynolds averaged Navier Stokes equation.Tirović[10] proposed a cooling to aerodynamic efficiency ratio and evaluated the performance of brake discs with radial vanes and circumferential pillars.

In summary, the academic community has conducted research on the performance of train brake discs using different evaluation indicators. In this article, the heat dissipation rib cross-sectional shapes and layout methods are compared and analyzed, and a new evaluation indicator and the pump wind-heat dissipation power ratio are proposed to study the optimal brake disc heat dissipation rib cross-sectional shape and layout method.

2. Research on the cross-sectional shape of the brake disc cooling fins

In this paper, the flow field characteristics and the pump-airflow to heat dissipation power ratio of four types of brake disc fin cross-sections are analyzed: diamond, square, ellipse, and circle, and the optimal cross-section for brake disc fins is obtained. The four types of brake disc fin cross-sections are shown in Figure 1, and each has a consistent characteristic scale. The diameters of the circular fin, the lengths of the minor axis of the elliptical fin, and the diagonal lengths of both the diamond and square fins are all 15 mm.

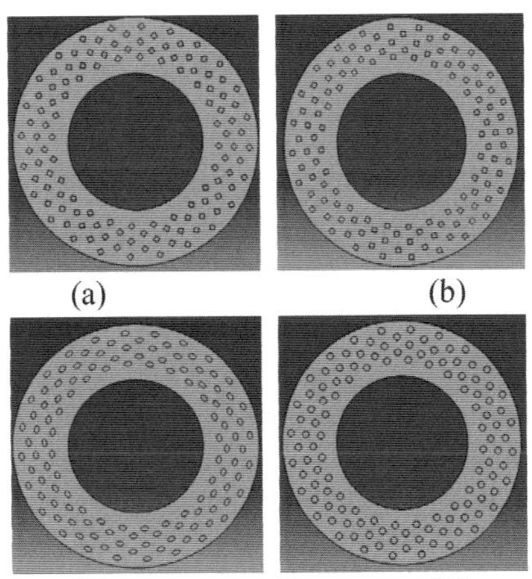
(a) (b)
(c) (d)

Figure 1. The cross-sectional shape of the cooling fins

The pump air-cooling power ratio is defined as the ratio of the brake disk pump air power to the cooling power, expressed as in Equation (1). The definition of pump air power is shown in Equation (2).

$$\eta_v = \frac{Q_{con}}{Q_{areo}} \tag{1}$$

$$Q_{areo} = Fv \tag{2}$$

In the equation, η_v represents the ratio of the pump air-cooling power to the heat dissipation power; Q_{areo} represents the pump air-cooling power, with a unit of W; Q_{con} represents the heat dissipation power, with a unit of W; F represents the air resistance, with a unit of N; v represents the velocity, with a unit of m/s.

2.1 Steady-State Analysis of Brake Discs.

By setting the brake disc to a constant rotational speed, a comparative analysis is conducted on the air resistance, the pump power, and the heat dissipation power of brake discs with different heat dissipation rib cross-sectional shapes during steady-state rotation. The pump air-cooling power to heat dissipation power ratio for each heat dissipation rib cross-sectional shape of the brake disc is obtained.

The air resistance experienced by brake discs with different heat dissipation rib cross-sectional shapes as a function of speed is shown in Figure 2. From the figure, it can be seen that the resistance experienced by the heat dissipation ribs increases in the following order: ellipse, circle, diamond, and square.

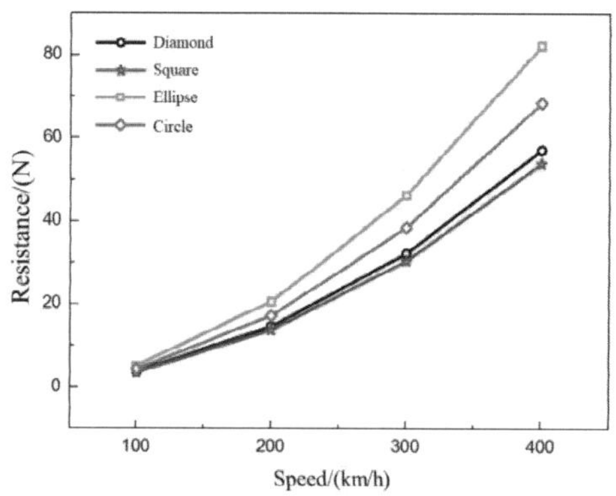

Figure 2. The different shapes of cross-sections correspond to different levels of air resistance

The pressure field of brake discs with different heat dissipation rib cross-sectional shapes is shown in Figure 3 when the incoming flow velocity is 100 km/h. Figures 3(a)-(d) correspond to brake discs with heat dissipation rib cross-sectional shapes of circle, ellipse, square, and diamond, respectively.

As shown in Figure 3(a), separation occurs on the upper and lower sides of the cylinder, but the overall wake width is small, and the negative pressure zone is small, resulting in reduced pressure drag. As shown in Figure 3(b), the elliptical configuration is similar to the circular configuration, but its separation is further back, with a shorter and narrower wake and relatively smaller pressure drag. In the square configuration (c), the area of the negative pressure zone in the wake is significantly larger, the square cylinder hinders the normal flow of the fluid and separation occurs at the upper and lower edges of the cylinder on the windward side, and the large separation angle after separation leads to a wider negative pressure zone in the wake, increasing pressure drag. In the diamond configuration (d), the heat dissipation ribs similarly obstruct the normal flow of the fluid and separation occurs at the upper and lower edges of the cylinder on the windward side, but the separation angle between the flow direction after separation and the horizontal direction is smaller than that in the square configuration, resulting in a narrower wake than in the square configuration. Therefore, during the steady rotation of the brake disc, the elliptical heat dissipation rib experiences the least air resistance.

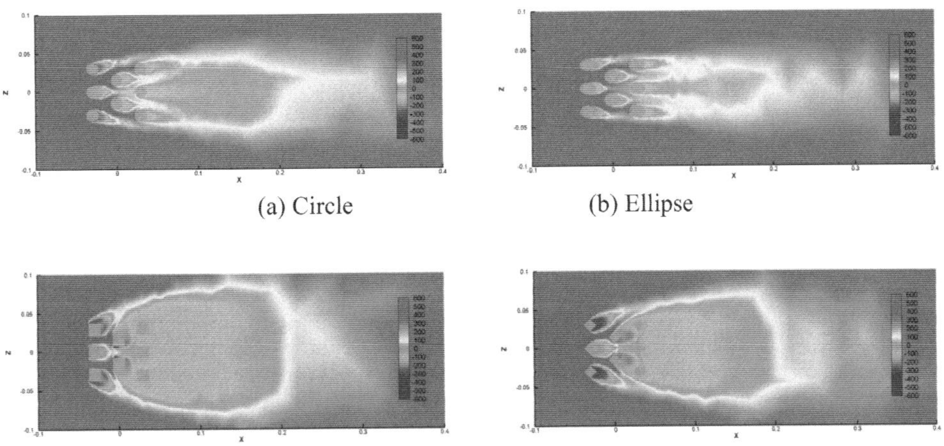

(a) Circle (b) Ellipse

(c) Square (d) Diamond

Figure 3. Pressure field corresponding to different cross-sectional shapes

The heat dissipation power of different section shapes of heat dissipation fins is shown in Figure 4. As can be seen from Figure 4, with the increase of the incoming flow velocity, the volume of airflow passing through the space per unit of time increases, taking away more heat and increasing the heat dissipation power. Under the same velocity condition, the heat dissipation power of the elliptical section is the highest, followed by the circular section, and the heat dissipation power of the angular section is the lowest.

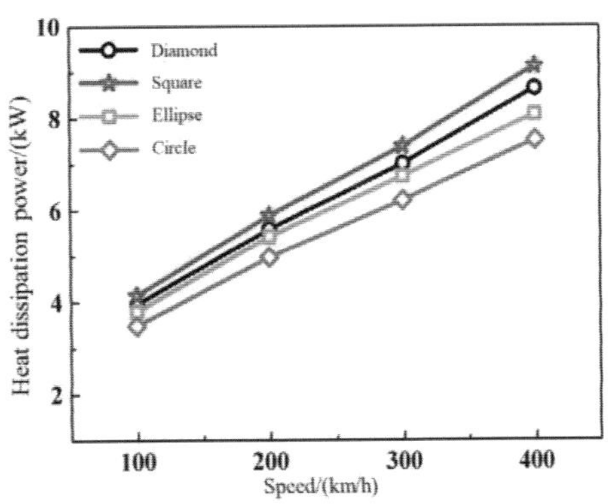

Figure 4. The different shapes of cross-sections correspond to different levels of heat dissipation power

Using Equation (1), the pump wind-heat dissipation power ratio can be obtained. The calculated results of the pump wind-heat dissipation power ratio for different heat dissipation fin cross-sectional shapes under different inflow velocities are shown in Figure 5. As can be seen from Figure 5, the elliptical heat dissipation fin has a larger heat dissipation power while maintaining a lower resistance, making it the optimal cross-sectional shape in steady-state calculations.

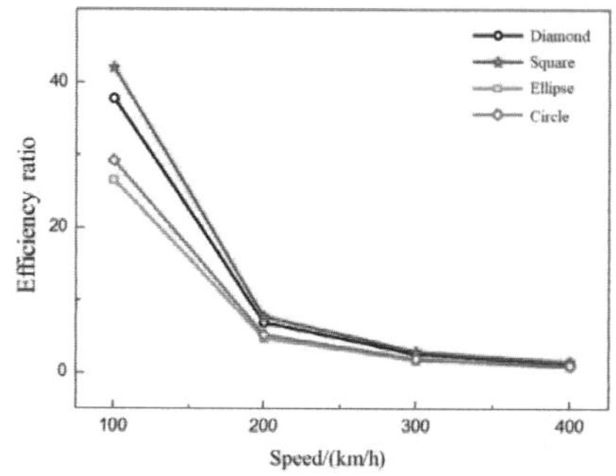

Figure 5. The power ratio of different cross-sectional shapes varies

2.2 Analysis of the transient flow field and pump air cooling performance of elliptical cylinder with heat dissipation fins

During braking conditions, the rotational speed of the brake disc gradually decreases. Therefore, based on steady-state analysis, the pump air cooling performance of elliptical heat dissipation fins on the brake disc is analyzed under transient conditions. Keeping the cross-sectional area of the heat dissipation fins constant, different aspect ratio elliptical heat dissipation fin structures are designed. The heat dissipation fins are arranged in a single circumferential direction on the disc, with the a-axis of the elliptical fins coinciding with the circumferential mirror direction and the b-axis coinciding with the tangential direction. The elliptical heat dissipation fin ratio is defined as b/a. Four different operating conditions were designed with elliptical heat dissipation fin ratios of 1:1.5 to 1.5:1, as shown in Table I.

TABLE I. Calculation Conditions

Work Condition No.	***Elliptical Heat Dissipation Fin Ratio***	***Radial Dimension a/mm***	***Tangential Dimension b/mm***	***Elliptical Cross-sectional Area mm²***
1	1： 1.5	24.495	16.33	400.0034
2	1： 1	20	20	400
3	1.25： 1	17.9	22.37	400.423
4	1.5： 1	16.33	24.495	400.0034

2.2.1 The air resistance experienced by brake discs with different elliptical heat-dissipating rib proportions

The relationship between the resistance and resistance moment of each part of the brake disc and the proportion of elliptical heat-dissipating ribs is shown in Table II. By comparing the resistance data calculated with different proportions of elliptical heat-dissipating ribs, it is found that as the proportion increases, the resistance increases and the resistance moment decreases, with a more significant change in the resistance moment.

TABLE II. The relationship between the resistance and resistance moment of each part of the brake disc and the proportion of elliptical heat-dissipating ribs

Work Condition No.	***Resistance***		***Resistance Moment***	
	Rib	***Total***	***Rib***	***Total***
1	23.76	165.31	9.41	12.86
2	27.03	169.50	5.91	9.34
3	30.41	172.81	4.29	7.74
4	34.40	176.91	2.85	6.29

At a flow velocity of 400 km/h, the pressure field distribution of brake discs with different proportions of elliptical heat-dissipating ribs is shown in Figure 6. As can be seen, for the windward side, as the proportion increases, the angle formed by the separation of the flow on the upper and lower sides of the cylinder increases, causing the negative pressure region of the wake to gradually increase, the pressure difference between the windward and leeward sides of the elliptical ribs increases, and the resistance increases. For the rotating direction, the angle formed by the separation of the flow in the circumferential direction of the elliptical cylinder decreases, causing the negative pressure region of the wake to gradually decrease, the pressure difference between the windward and leeward sides of the elliptical ribs decreases, and the resistance moment decreases.

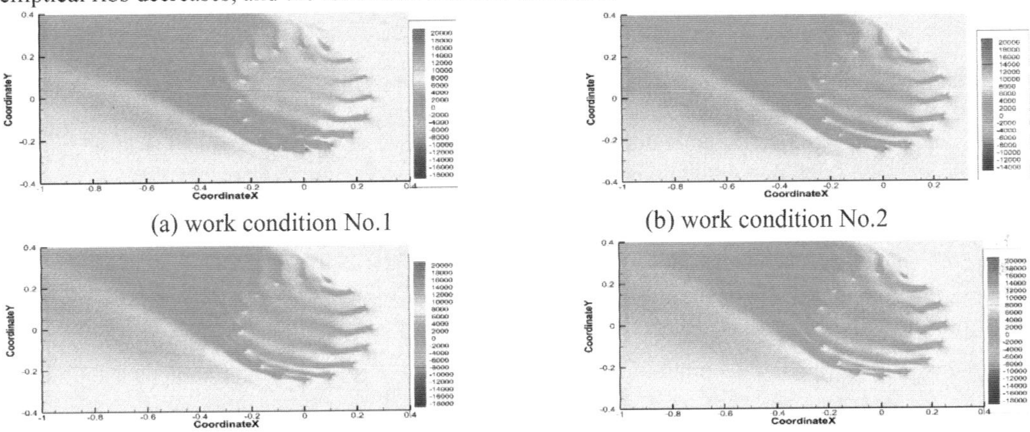

(a) work condition No.1 (b) work condition No.2

(c) work condition No.3 (d) work condition No.4

Figure 6. The pressure field corresponds to different working conditions

2.2.2 Different elliptical cooling fins proportions affect the heat dissipation of brake discs

The simulation results of brake disc heat dissipation power are shown in Table III. As can be seen, with the increase in the proportion of elliptical cooling fins, the heat dissipation power shows a decreasing trend.

TABLE III. The relationship between the heat dissipation power of the brake disc and the proportion of elliptical cooling fins

Work Condition No.	***Heat dissipation power***	
	Cooling Fins	***Total***
1	4183.15	47500.22
2	4192.93	46776.62
3	4199.17	46674.90
4	4202.39	46699.19

2.3 Pumping air to heat dissipation power ratio varies with different proportions of elliptical cooling fins

Using the given formula for the pumping air to heat dissipation power ratio, the pumping air to heat dissipation power ratio for different proportions of elliptical cooling fins is calculated, and the results are shown in Table IV. It can be seen that under the transient operation of the brake disc, Condition 2 (circular cooling fins) shows the smallest resistance and the highest heat dissipation power, making it the optimal cross-sectional shape.

TABLE IV. The relationship between the pumping air to heat dissipation power ratio and the proportion of elliptical cooling fins of the brake disc

Work Condition No.	***Heat dissipation power***	***pumping air power***	***Heat dissipation power to pumping air power ratio***
1	47500.22	21492	2.210
2	46776.62	21101	2.217
3	46674.90	21082	2.214
4	46699.19	21184	2.204

3. Research on the cross-sectional shape of the brake disc cooling fins

As revealed in the previous studies, circular cooling fins have a better-pumping air to heat dissipation power ratio. In this section, the arrangement method of circular cooling fins is studied. The comparison study is conducted between the parallel arrangement and the cross arrangement of cooling fins. The parallel arrangement is shown in Figure 7(a) while the cross arrangement is shown in Figure 7(b). The optimal arrangement method of cooling fins is obtained by measuring the pumping air to heat dissipation power ratio.

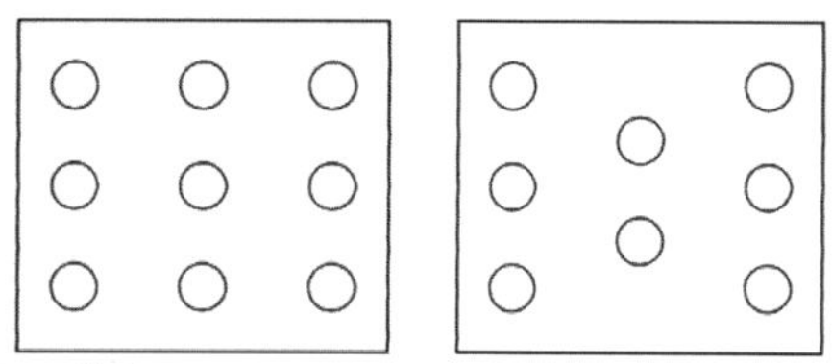

(a) Parallel arrangement (b) Cross arrangement

Figure 7. Schematic diagram of cooling fins arrangement

3.1 Analysis of flow field and pumping air heat dissipation performance for different cooling fin arrangements

3.1.1 Air resistance experienced by brake discs with different cooling fin arrangements

The air resistance experienced by cooling fins with different arrangements is shown in Figure 8. From Figure 8, it can be seen that the air resistance of the cross arrangement is slightly higher than that of the parallel arrangement, and the maximum air resistance of the cross arrangement is only 1.24 N·m greater than that of the parallel arrangement. Therefore, the air resistance experienced by the brake discs with these two cooling fin arrangements is relatively small.

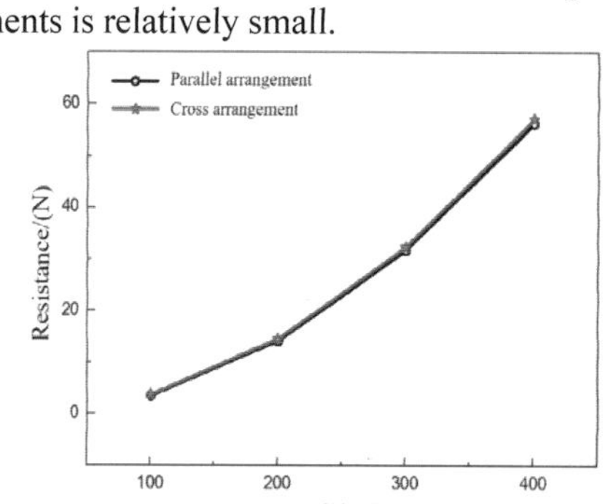

Figure 8. Different arrangements of cooling fins on brake discs correspond to different air resistance

By solving the pressure field and velocity field distributions corresponding to the two types of arrangement, the pressure cloud map and velocity cloud map shown in Figures 9 and 10 are obtained. From the cloud maps, it can be seen that due to the influence of the wake of the front row of cooling fins, the wake width of the parallel arrangement is significantly smaller than that of the cross arrangement, and the pressure on the windward side is smaller, resulting in a smaller pressure difference between the front and rear, and therefore a relatively small resistance for the parallel arrangement.

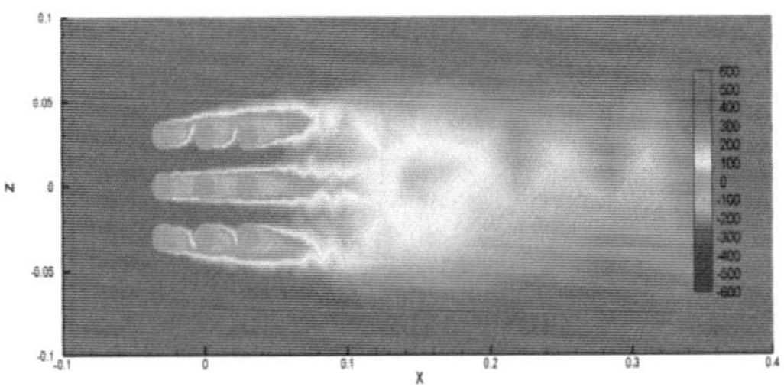

(a) Parallel arrangement

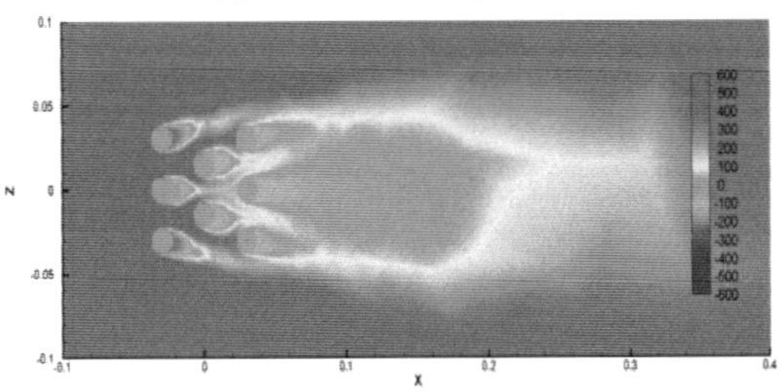

(b) Cross arrangement

Figure 9. Pressure field corresponding to different cooling fins arrangements

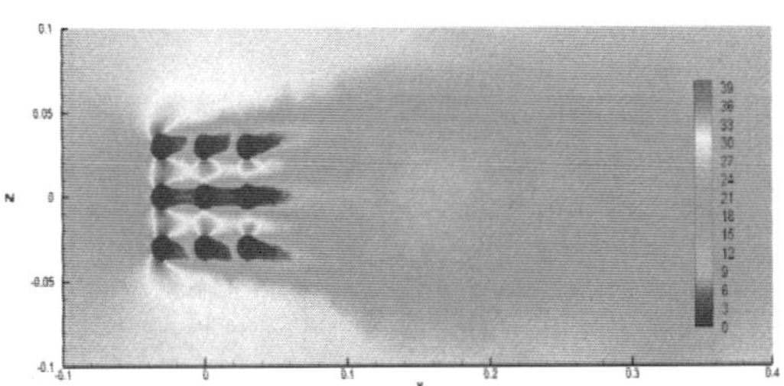

(a) Parallel arrangement

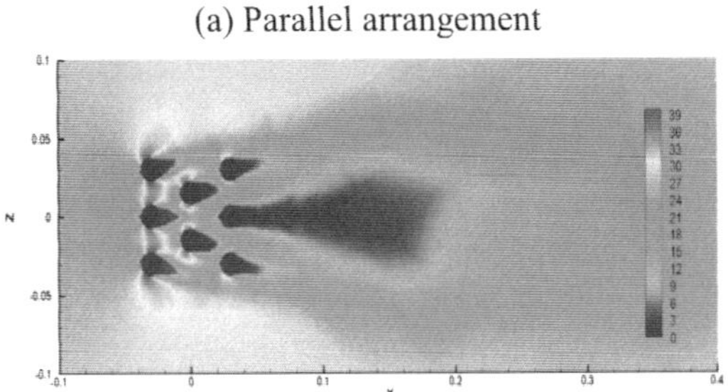

(b) Cross arrangement

Figure 10. Velocity field corresponding to different cooling fins arrangements

3.1.2 Different arrangements of cooling fins correspond to the heat dissipation power of the brake disc

Different arrangements of cooling fins correspond to the heat dissipation power of the brake disc, as shown in Figure 11. As can be seen, under the same speed conditions, the heat dissipation power of the cross arrangement is higher than that of the parallel arrangement, but the difference between the two decreases as the speed increases. When the cooling fins are arranged in a cross pattern, the fluid flows in a curved channel between the fins, where the flow alternates between contraction and expansion, resulting in more severe turbulence and more efficient heat exchange between the airflow and the cooling fins, leading to higher heat dissipation power. However, as the speed increases, the airflow is carried away before it can fully develop into a vortex, reducing the difference in heat dissipation power between the cross-arrangement and the parallel arrangement. Therefore, the cross arrangement of heat dissipation fins is superior to the parallel arrangement in terms of heat dissipation power for brake discs.

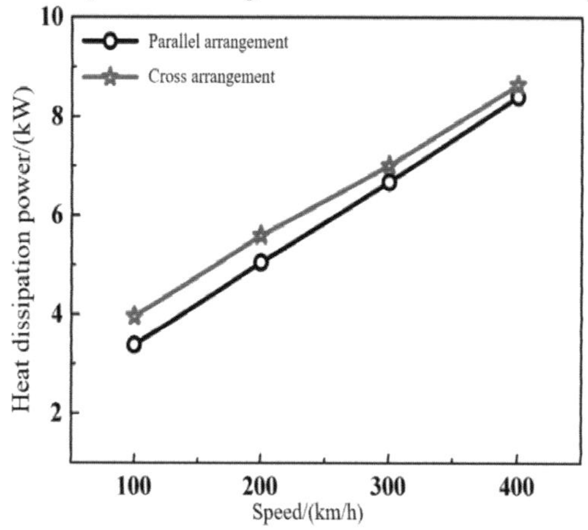

Figure 11. The corresponding heat dissipation power for parallel and cross arrangement

3.1.3 Pumping power to heat dissipation power ratio for different cooling fins arrangements

Using the pumping power to heat dissipation power ratio formula provided in the previous text, the corresponding ratio for different cooling fins arrangements is calculated. As shown in Figure 12, it can be seen that the pumping power to heat dissipation power ratio for different speed cross arrangements of the brake disc are all better than those for parallel arrangements.

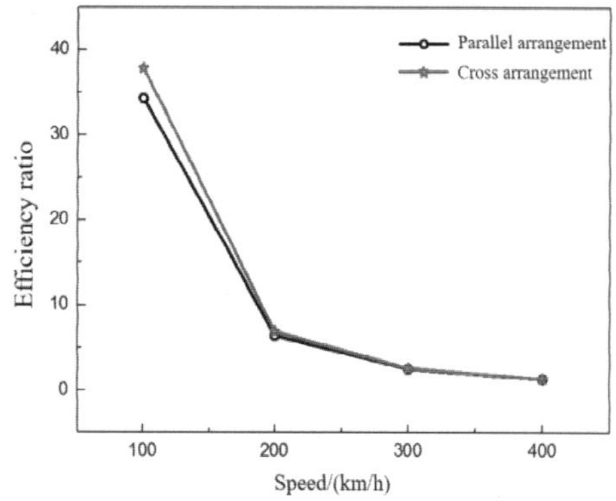

Figure 12. Pumping power to heat dissipation power ratio for parallel and cross cooling fins arrangement

4. Conclusion

In this article, the effects of different cooling fin shapes and arrangements on the pumping and heat dissipation characteristics of the brake disc were compared and analyzed through simulation, and a new evaluation index was proposed, namely the pumping power to heat dissipation power ratio. The optimal cooling fin cross-sectional shape and arrangement of the brake disc were studied, and the following conclusions were obtained:

(1) Under steady-state conditions, the elliptical cooling fin brake disc has better pumping and heat dissipation performance. Under transient conditions, the circular cross-sectional cooling fin has the highest pumping power to heat dissipation power ratio and optimal pumping power loss and heat dissipation power, making it the optimal cross-sectional shape.

(2) The resistance of the cooling fin with the cross arrangement is slightly higher than that with the parallel arrangement, but for heat dissipation performance, the cross arrangement shows better results than the parallel arrangement. Considering the pumping power to heat dissipation power ratio, the comprehensive performance of cross arrangement is better than that of parallel arrangement.

Acknowledgment

The authors gratefully acknowledge the financial support by Research on the wheel-rail adhesion mechanism and adhesion improvement under large slip conditions- the National Science Foundation of China under Grant numbers 52072266 and Simulation of wheel-rail relationship and modeling of anti-slip control under high-speed and ultra-low adhesion conditions (P2021J038).

References

[1] Zuo J. Y., Luo Z. J. (2014) Air-pumping Effect Analysis for Brake Disc of High-speed Train. Journal of Traffic and Transportation Engineering, 14: 34-40+74.
[2] Zuo J. Y., Liu J. L., Hu G., Gu Y. H. (2018) Analysis on the Structural Parameters of Train Brake Disc Based On Transient Temperature Simulation. Machine Design and Manufacturing Engineering, 47:13-18.
[3] Ma S. Q., Zhao W. Z., Xie S. M., Wan C. Y. (2003) Numerical Simulation of Thermal Mechanical Coupling Process of Train Brake Discs. Machine Design and Manufacture, NO.4:73-74.
[4] Zhang Y. C. (2022) Optimization Design of Heat Dissipation Ribs for High-speed Train Brake Discs.China Science and Technology Information, NO.10:115-117.
[5] Suo R. Q., Shi X. L. (2022) Temperature Field and Stress Field Distribution of Forged Steel Brake Disc for High Speed Train. Jordan Journal of Mechanical & Industrial Engineering, 16:113-121.
[6] Kravchenko K., Hauser V., Gorbunov M. (2019) Estimating the Brake Disc Air Resistance By Scaled Test Rig and By Computer Simulation. Transportation Research Procedia, 40:737-744.
[7] Zuo J. Y., Luo Z. J. (2014) Conjugate heat transfer study on a ventilated disc of high-speed trains during braking. Journal of Mechanical Science and Technology, 28: 1887-1897.
[8] Wang Y. G., Zhou Q., Qiao Q. F., Ma L. J. (2016) Heat dissipation performance simulation on brake disk of high-speed train based on aerodynamics. Computer Aided Engineering, 25: 25-28.
[9] You J. L., Chen Y., Yang Z. G. (2015) Influence of the Wheel Rotation on Underbody Flow and Aerodynamic Forces of High Speed Train. Procedia Engineering, 126:399-404.
[10] Tirović M. (2009) Energy thrift and improved performance achieved through novel railway brake discs. Applied Energy, 86: 317-324.

Design Method of the Civil Aircraft Cargo Compartment Interiors Based on Resist-flame Burnthrough

Huan Zhang

5188 Jinke Road, Pudong New Area, Shanghai

zhanghuan@comac.cc

Abstract. This paper aims to conduct research on interior design methods based on resist-flame burn-through. The paper first analyzed the acceptance criterion of the oil burner test according to part III of Appendix F, part 25, [1]. Then it elaborated on various design methods of cargo compartment liners combined with the previous design and test experience, including the design method of material, opening features, lap/butt joint features, and corner butt joint features [2]. These design methods can effectively improve the ability to resist-flame burn-through of cargo compartment liners.

1. Introduction

The cargo compartment, which is under the passenger compartment in civil aircraft, is generally defined as a C-class cargo compartment. Breaking out of the fire in the cargo compartment could threaten the lives of passengers in the passenger compartment. Thus, the airworthiness regulation demands that the cargo compartment liner should have the capability to contain a fire, in other words, meet the oil burner test requirements of §25.855(c). The main cargo compartment of the freighter is generally defined as an E-class cargo compartment. The airworthiness regulation does not demand E-class cargo compartments to meet the oil burner test requirements. However, this may conduct unsafe conditions based on the experience of similar aviation products. So many main cargo compartments of freighters are additionally demanded to meet the oil burner test requirements in §25.855(c), such as Boeing 777F, Boeing 767-300F, and so on.

For the cargo compartment liner, the conditions of the oil burner test are very harsh. Because of the high temperature and the long burning time during the test, the conventional design of the cargo compartment liner probably could not meet the requirement of resist-flame burn-through. So, the characteristics of this test should be taken into full consideration in the early stage of design and conduct the liner design based on resist-flame burn-through. This article carries on research on the design of the material and the feathers based on existing design and test experiences.

2. Analysis of Resist-flame Burnthrough Acceptance Criterion

The oil burner test applicable to cargo liners contains two pass/fail criteria in part III of Appendix F, part 25. Firstly, the liner successfully resists penetration by the burner flame for five minutes. Secondly, the temperature on the backside of the ceiling liner test panel does not exceed 400°F when measured 4 inches above the test panel. The first criterion is fairly straightforward in that the liner should remain intact throughout the test. The second criterion is designed to screen out materials that would arrest flame but would not otherwise mitigate the effects of the fire [3].

Meanwhile, not only the cargo liner itself should pass the oil burner test, but also the design feathers on or between the liners, such as the equipment on the liners, and the joints between the liners [4].

3. Material Design Method Based on Resist-flame Burn-through

Cargo interiors are the liners used to separate the cargo compartment and the structure. Cargo interiors mainly consist of composite materials and metallic materials. Plate-type part of the large area usually chooses composite plate, and the part used to install and fix other parts usually chooses metallic material [5].

On the choice of composite materials, the interior cargo panels generally should resist the burn-through of the fire. So, when choosing the edge banding adhesive, the type of adhesive that takes more time to be ignited and is not easy to melt and produce drops is preferred. The bigger the size of the edge banding is, the more adhesive the interior panel uses, so the panel is more easily ignited or burns for more time. Therefore, interior cargo panels can get a better ability to resist the burn-through of fire by reducing the size of edge banding [6].

On the choice of metallic materials, the melting point of aluminum alloy is generally not more than 660°C, and the flame's temperature in the oil burner test can reach 927 °C. So, different from the usually chosen aluminum alloy in conventional passenger compartment interior design, the metallic parts in the visible side of the cargo compartment are usually used to bordure the edge of the panels or to fix two panels in the lap joints [7][8].

The layers that press the edge or cover the gaps usually choose decorative laminates or titanium alloy.

The accessories used in the cargo compartment, such as silicon rubber gaskets, are usually very easy to be ignited. So, accessories are not recommended to be used on the ceiling. If the ceiling needs to be sealed, this can be realized by pasting sealing tape on the compartment side [9].

4. Opening Features Design Based on Resist-flame Burnthrough

Common opening feathers in the cargo compartment can be divided into the following types:

- Opening feathers for equipment installation. This type of opening feather is used to install equipment, such as smoke detector opening feathers.
- Opening feathers for flexible shaft passing. The opening provides a passageway for the flexible shaft, such as the cargo door actuator flexible shaft opening feather.
- Opening feathers for wire penetration. The wire of the equipment penetrates through the opening.
- Gap opening feathers. The gap usually forms because interior panels cannot match each other completely, such as the confluence of three panels.

4.1. Design method of equipment installation opening feathers

For equipment installation opening feathers, firstly, the equipment itself or its protective cover should block the opening, arrest the flame, and resist the burn-through of the flame.

For equipment that does not have protective covers, the region, such as the exterior shell, should have the ability to resist-flame burn-through. Figure 1 presents the opening feather of the cargo door light. To resist the burn-through of flame, the material of the light shell could choose titanium alloy.

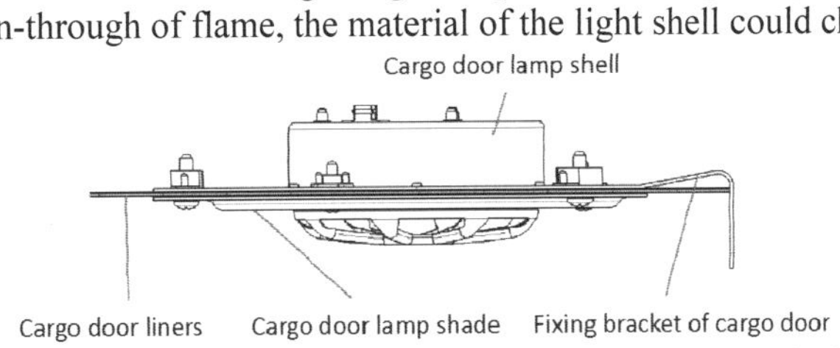

Figure 1. Opening Feather of Cargo Door Light (Typical).

To avoid the impact on the goods or the install the equipment, many pieces of equipments have protective covers. And to meet the requirement of resist-flame burn-through, a protective cover can be designed in particular. The material of the protective cover usually chooses composite laminated plate, and the cover usually connects with the interior panel by fasteners through its turnup. If the protective cover cannot resist the burn-through of flame, the following design methods can be taken into consideration to improve its ability to resist flame burn-through.

4.1.1. Increase the amount of overlap between the cover turnup and the interior panel.
According to the experience of past tests, when the protective cover and the interior panel are both made of the composite laminated plate, if the overlap is not less than 30 mm, the design feather is more likely to pass the test. And the ability of resist-flame burn-through can be improved by increasing the amount of overlap between the cover turnup and the interior panel.

4.1.2. Add the turnup pressing design of mental blank holder.
A mental blank holder can be added on the visible side of the interior panel or the invisible side or adding two mental supports synchronously on both sides. For example, the protective cover of the fire nozzle is made of a laminated plate. In the actual test, the turnup and the opening edge where the fire nozzle pass through can be burned through easily. Two mental blank holders, as Figure 2 shows, can be added to press the turnup of the cover and the opening edge.

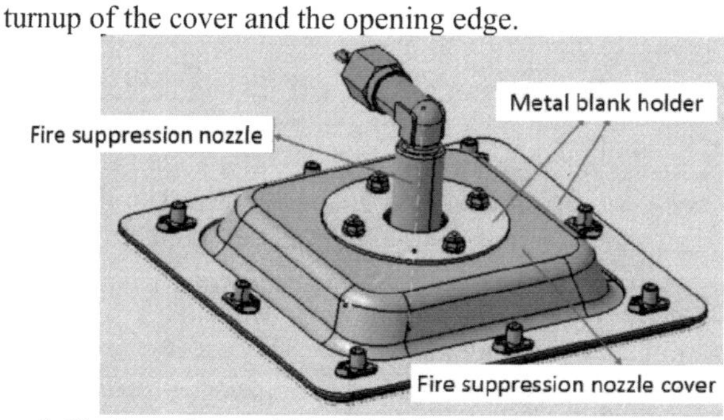

Figure 2. Design of mental blank holder for fire suppression nozzle.

4.1.3. Increasing the number of fasteners.
If the turnup and the edge of the opening are easy to become deformed because of heating, increasing the number of fasteners can improve the ability to resist flame burn-through. For example, the protective cover is fixed by 4 fasteners with the interior panel at the 4 corners of the cover. Another 4 fasteners are added at the 4 sides of the turnup additionally to improve the resist-flame burn-through ability of the turnup, as Figure 2 shows.

4.2. Design method of opening feathers for flexible shaft passing
The most typical opening feathers for flexible shaft passing is an oblong opening. For example, there is a 300 mm long and 44 mm oblong opening on the ceiling panel in one freighter of a certain type, which the flexible shaft of the main cargo compartment door actuator passes through.

For opening feathers for flexible shaft passing, to meet the requirements of resist-flame burn-through, a protective cover can be designed on the visible or invisible side of the cargo compartment. The material usually chooses decorative laminates. If the kinematic envelope of the shaft is on the visible side of the cargo compartment, the protective cover should be on the other side (the invisible side of the cargo compartment), and vice versa. A typical design of the protective cover in one freighter of a certain type can be seen in Figure 3. The oblong opening is on the plate, fixed to the ceiling panel. As the kinematic envelope of the shaft is on the visible side, the protective cover is designed on the back side of the plate. The cross-section shape of the flexible shaft exit is semi-circular, and the exit is designed as a tunnel. The longer the tunnel-shaped exit is, the more difficult the flame is to escape from the exit. The position, shape, height, and other dimensions of the protective cover should take the minimum turning radius of the shaft into consideration, and the movement of the shaft should not be affected.

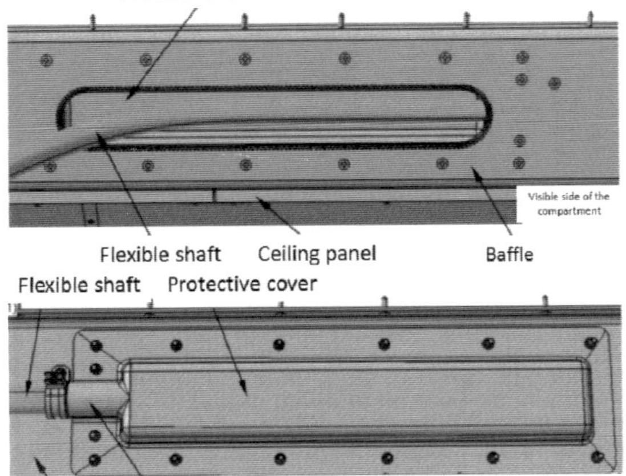

Figure 3. Typical design of the protective cover.

4.3. Design method of opening feathers for wire penetration

For opening feathers for wire penetration, according to AC, holes in ceiling panels not larger than 0.375 inches in diameter are used to provide access to smoke detector sampling ports. So, when designing the opening for wire penetration, we should make sure that the maximum diameter of the opening shall not exceed 3/8 inch. However, wire penetrations through a non-ceiling cargo liner do not require to be limited by 3/8 inch if all of the following are met.

- They are required for the functionality of cargo systems.
- The equipment the wiring serves is in a shroud made of a material that meets §25.855(c).
- The penetrations are located in the vertical or lower surface of the equipment shroud, as Figure 4 shows.

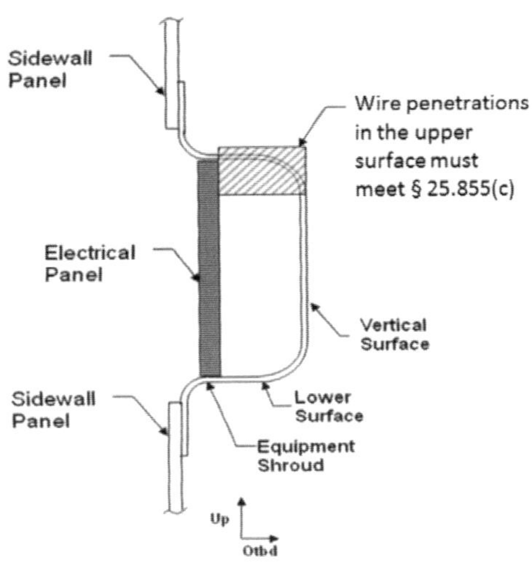

Figure 4. Equipment wire penetrations.

4.4. Design method of gap opening feathers

According to AC, gap opening feathers as follows can meet the requirements of §25.855(c):

- Sidewall liner gaps should be no larger than 0.25 inch around system penetrations (e.g., wire, pipe), or gaps should be no larger than 0.15 inches wide by 4 inches long and are typically associated with splices in substructure exposed between butt joint liners.
- Ceiling liner gaps should be no larger than a 0.125-inch diameter circle.

During the designing of the interior liners, the dimensions of the gap opening could be controlled within the range mentioned above in (1) and (2). If the dimensions of the gap opening do not meet the requirements, additional design is needed to reduce or eliminate the gaps, such as increasing the design of the covering strip or covering.

5. Lap/Butt Joint Features Design Based on Resist-flame Burn-through

5.1. Design method of lap joint feature

The connection mode of two laminate panels in the cargo compartment is generally a lap joint. As the laminate panel is very thin, it is easy to warp when being burned at high temperatures. So, when conducting the design of lap joint features, according to the experience, the lap width is generally not less than 30 mm, and the fastener pitch is generally not more than 100 mm. If the lap width is too small or the fastener pitch is too big, the design of the depression strip generally needs to be added. Figure 5 is the typical design.

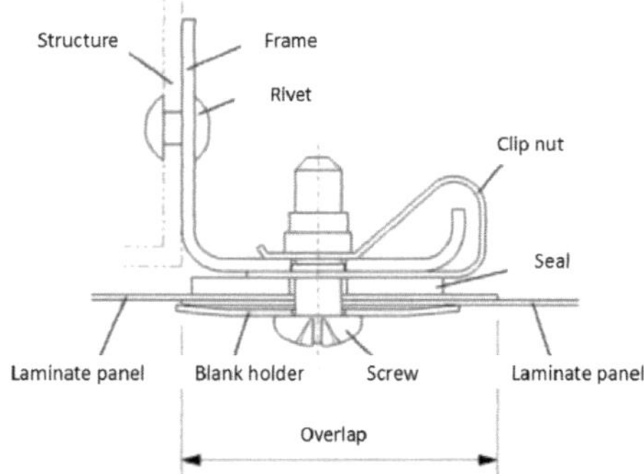

Figure 5. Typical lap joint.

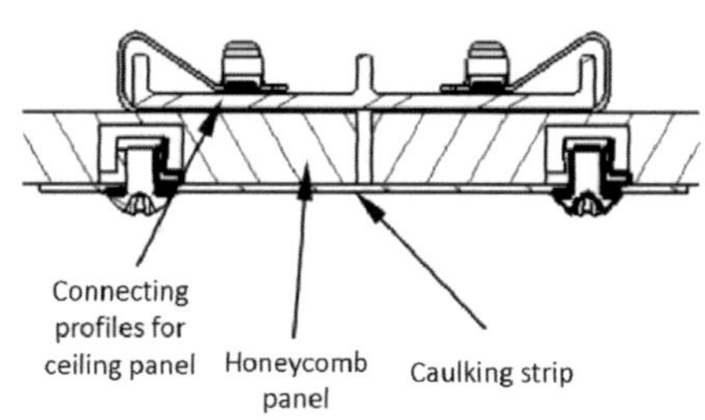

Figure 6. Typical butt joint.

5.2. Design method of butt joint feature

The connection mode of two honeycomb panels in the cargo compartment is generally a butt joint. And a mental structural frame or support for connection or fixation is generally designed behind the seam. If the frame or support is on the invisible side of the cargo compartment, the material could choose aluminium alloy. The width of the seam is generally not more than 3 mm to avoid the burn-through of the flame. If the width of the seam is too big, the design of the covering strip generally needs to be added to cover the seam on the visible side of the cargo compartment. In addition, the material of the strip could choose the laminate panel or the titanium alloy plate. Figure 6 is the typical design.

6. Corner Butt Joint Features Design Based on Resist-flame Burn-through

According to the test experience, the width of the seam in the corner butt joint feature is generally not more than 7 mm to avoid the burn-through of the flame. The measurement method of the seam width is shown in Figure 7 [10].

If the width of the seam is too big in the corner butt joint, an L-shaped covering strip or strip of other shapes could be designed to cover the seam. The material of the strip could choose the laminate panel or the titanium alloy plate. Figure 8 is the typical design.

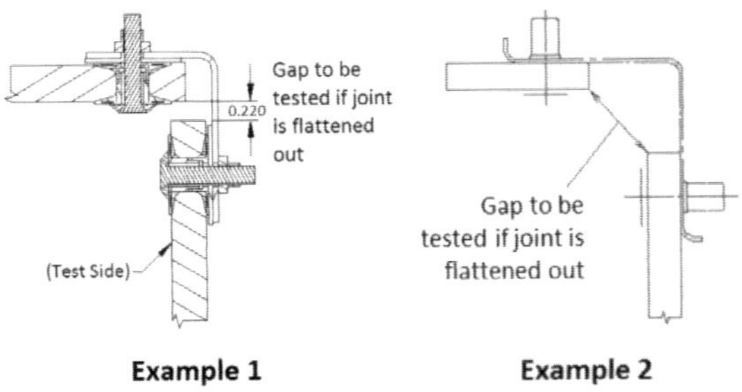

Figure 7. Gap measurements.

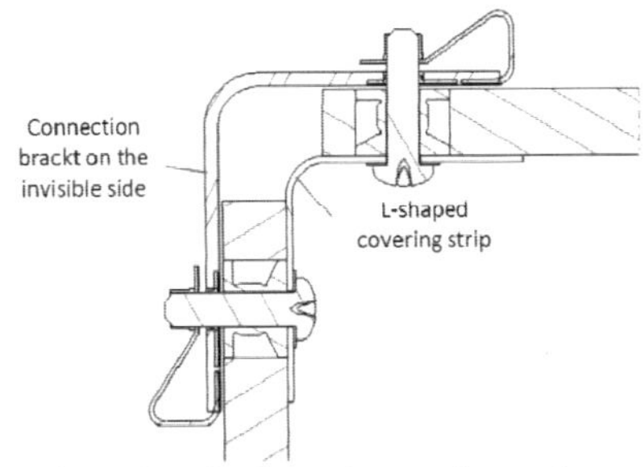

Figure 8. Design of an L-shaped covering strip

7. Conclusions

Based on previous design experience and test experience, this thesis conducted research on the interior design method based on resist-flame burn-through. It includes the design method of material, the design method of Opening Features, the design method of Lap/Butt Joint Features, and the design method of Corner Butt Joint Features. And specific design schemes are given based on resist-flame burn-through. These design methods can effectively improve the ability to resist-flame burn-through of cargo compartment liners.

References

[1] Civil Aviation Administration of China. CCAR25-R4 Civil Aviation Regulations of China Department 25：Airworthiness Standards for Transport Aircraft [S]. Beijing: Aviation Administration of China, 2011.

[2] FAA. Transport Airplane Cabin Interiors Crashworthiness Handbook [S]. No: 25-17A. Federal Aviation Administration, 05. 18. 09.

[3] FAA. Aircraft Materials Fire Test Handbook, Revision 3[S].DOT/FAA/TC-17/55, Federal Aviation Administration, July 2019.

[4] Kuang Lili. Airworthiness Requirements of Fire Prevention and Compliance Verification Test Method for Transport Airplane Interior Materials[J]. Civil Aircraft Design & Research, 2018, 128(01): 19-23.

[5] Li Pengyang. Discussion about Flammability Airworthiness Verification Method of Civil Aircraft Composite Material[J]. Civil Aircraft Design & Research, 2022, 147(04): 108-112.

[6] Niu Fangxu, Sun Chaoming, Zhu Xiudi, Yang Xiaolin, He Jing. Preparation and Performance Research of Nomex-Honeycomb Sandwich Cargo Floor Panel for Aircraft[J]. Fiber Composites, 2020, 04: 100, 110-113.

[7] Xve Yangwu. Study on Combustion Characteristics of Civil Aircraft Cargo Cabin Panel Materials[D]. Si Chuan: Civil Aviation Flight University of China, 2021.

[8] Zhang Zhang, Cha Meng. Research on Application of Flame-retardant Composite in Plane Cabin[J]. New Chemical Materials, 2021, 49(11): 252-257.

[9] Huang Song. Study on Combustion Properties of Typical Interior Materials of Aircraft under Low Ambient Pressure[D]. Si Chuan: Civil Aviation Flight University of China, 2019.

[10] Timothy Marker. Burnthrough Test Procedures for Cargo Liner Design Features[S]. DOT/FAA/CT-TN88/33, Federal Aviation Administration Technical Center, September 1988.

Optimization of CNC Milling Parameters Based on Tool Life and Process Carbon Emission Prediction Models

Wangyuan Yi, Ruixue Yin

School of Mechanical Engineering, Guizhou University, Guiyang City, China 550025

e-mail: ywygzu@126.com e-mail: yinruixue@sina.com

Abstract. Aiming at the problems of short tool life and high carbon emission in the process of CNC milling, a multi-objective optimization method of process parameters based on the BP neural network and NSGA-II algorithm is proposed. Based on the data sets generated by CNC milling under different process parameters, with tool life and process carbon emissions as outputs and different process parameter combinations as inputs, a BP neural network-based tool life and process carbon emissions prediction model is constructed. Taking the maximum tool life and minimum process carbon emission as the optimization goals, the NSGA-II process parameter multi-objective optimization main model is constructed. The model is then used as the objective function of the main model and optimized to obtain the Pareto optimal solution set. The optimal solution decision is made on the Pareto optimal solution set using the Topsis method, and the optimal process parameter combination is obtained. The optimization results show that the method can accurately predict the tool life and process carbon emission of CNC milling. It can also effectively optimize the process parameters, which provides a new method for the low-carbon optimization of CNC milling parameters.

1. Introduction

In the process of CNC milling, adjusting the process parameters can usually improve the processing efficiency of CNC milling. However, it often causes increased cutting heat, heat dissipation inequality phenomenon, speeds up tool wear, and reduces tool life, increasing the carbon emissions and cost of processing technology. Therefore, it has attracted the attention of scholars on how to determine the relationship between tool life, carbon emissions, and process parameters in the process of CNC milling and obtain the relatively optimal process parameters to achieve low cost, high efficiency, and low carbon emissions of CNC milling.

At present, domestic and foreign scholars have carried out a series of research on the optimization of process parameters. Rao et al. [1] established a multi-step process parameter optimization model for milling processing with machining time as the goal and obtained the optimal process parameters under constraints such as shank strength and deflection. Liu et al. [2] studied the relationship between high-speed dry-cutting hobbing process parameters and tool life, processing energy consumption, efficiency, and workpiece quality, which reduced the processing energy consumption and cost while prolonging the life. Zhang et al. [3] proposed a process parameter optimization method combining a neural network and genetic algorithm to solve the problem of tool wear and energy consumption in the side plate of micro-milling radial tire mold. Chen et al. [4] showed through high-speed milling processing experiments that reasonable process parameter values can consider production efficiency, cost, and quality. Mativenga et al. [5] established the minimum energy consumption criterion of machine tools and used the minimum energy consumption as the goal to optimize the three elements of cutting. Li et al. [6] established a multi-objective optimization model for CNC gear hobbing parameters with the lowest energy consumption and minimum processing time optimization goals. Chen et al. [7] optimized the process parameters for minimum energy consumption, minimum quality error, and maximum tool life for the high-speed dry hobbing process without cutting fluid and high cutting speed.

Among the above research on process parameter optimization, most are aimed at optimizing process parameters with processing time, processing efficiency, and processing quality as optimization goals. However, there are fewer studies on the prediction and optimization of tool life and process carbon emissions as process parameter optimization methods. Therefore, at the same time, the research on optimizing process parameters for CNC milling tool life and process carbon emissions is of great significance.

2. Optimization goals

In the CNC milling process, the setting of cutting parameters will affect the size of the milling force in the milling process, thereby affecting the tool life and the size of the carbon emissions [8]. In this paper, aiming at the optimization problem of CNC milling parameters, the cutting depth a_p, cutting speed, and feed f are selected as the optimization variables.

2.1Tool life

Tool life refers to the time a tool is put into service until it is actually scrapped. However, for different tool materials, workpieces, and cutting processes, the definition of tool life is not uniform. Therefore, one way to quantify the end of tool life is to set a maximum acceptable flank wear limit value (expressed in VB or VBmax). Therefore, the machining time used by the maximum wear limit value of the tool is used as the tool life (denoted as T).

2.2 Process carbon emissions

Process carbon emissions refer to the carbon emissions generated by CNC milling machines under different process parameters. Since CNC milling produces carbon emissions during machining, the main source is due to the consumption of electrical energy. Because energy consumption can be calculated by multiplying power by time, the resulting carbon emissions are calculated by multiplying energy consumption by the energy emission factor, as shown in Equation (1).

$$C = W \times F_e \tag{1}$$

where C means the process carbon emissions ($kgCO_2e$); W means the product of real-time power and machining time in the machining process of the machine ($kW \cdot h$); F_e means the Power Emission Factor ($kgCO_2e/kW \cdot h$).

3. Construction of BP neural network model

3.1 CNC milling experimental data set

This dataset is derived from [8][9]. In this experiment, the milling data under different parameter conditions are collected, as shown in Table 1. The collected data and the carbon emission calculation results according to Equation (1) are shown in Table 2.

The carbon emissions, tool life, and corresponding process parameters shown in Table 2 are data sources for BP neural network training and testing.

3.2 Generation of BP neural network prediction model

BP neural networks consist of multiple layers. For the determination of each level node, the nodes of the input layer and output layer are determined by the specific research problem. There is no standard selection rule for the selection of the number of hidden layer nodes.

Table 1. Experimental conditions

Number	a_p/(*mm*)	*f*/(*mm/rev*)
1	1.5	0.5
2	1.5	0.25
3	0.75	0.25
4	0.75	0.5

Table 2. Test data of CNC milling tool life and carbon emissions

Number	Number of passes	VB/ (*mm*)	a_p/ (*mm*)	*f*/ (*mm/rev*)	V_c/ (*m/min*)	*C*/ (*kgCO₂e*)	*T*/ (*min*)
1	10	0.7	1.5	0.25	200	2645.1544	57
2	14	0.55	0.75	0.5	200	2877.2317	72
3	16	0.55	0.75	0.25	200	2640.0978	81
4	17	0.44	1.5	0.5	200	2475.1648	48
5	9	0.81	1.5	0.5	200	2397.7014	46
…	…	…	…	…	…	…	…
94	23	0.76	0.75	0.25	200	3515.8097	105
95	15	0.65	0.75	0.5	200	2996.4853	74

Therefore, the selection range of the hidden layer nodes is determined according to the empirical Equation (2) [10].

$$h = \sqrt{m+n} + a \tag{2}$$

where h means the number of hidden layer nodes; m means the number of output layer nodes; n means the number of nodes in the input layer; a means an integer between 0-10. According to the optimization problem of CNC milling process parameters to be solved, a 3-layer BP neural network structure was constructed. Its main steps are as follows: ①Determine input and output parameters. According to the experimental data set of CNC milling under different operating conditions, the input parameters of the input layer are mainly CNC milling feed f, cutting depth a_p, cutting speed V_c, tool wear amount VB. The output parameters of the output layer are the process parameter optimization target, that is, the prediction target, tool life T and process carbon emission C. Therefore, it is determined that the input layer node of the BP neural network constructed in this paper is 4, and the output layer node is 2. ②Determination of hidden layer nodes. The interval selected by the hidden layer node is determined by Equation (2) to be 2-12. After multiple multi-objective predictions, it is found that when the number of hidden layer nodes is 10, the corresponding error is the lowest. ③Data normalization. To avoid the problems of slow network convergence and long training time caused by too large input data range, the function mapminmax is used to normalize the data. Therefore, the multi-objective prediction structure of the BP neural network is determined to be 4-10-2. The built BP neural network model is used to predict the target. The training number is set to 100, the learning rate is 0.12, and the minimum error of the training target is 1e-9. The prediction results are shown in Figure 1.

The results show that the coefficient of determination R^2=0.9750 for tool life prediction and R^2 =0.9065 for carbon emission prediction using BP neural network are used. The coefficient of determination of the prediction target is close to 1, and the error is within the allowable range. Therefore, the prediction accuracy of the BP neural network for tool life and carbon emissions meets the demand.

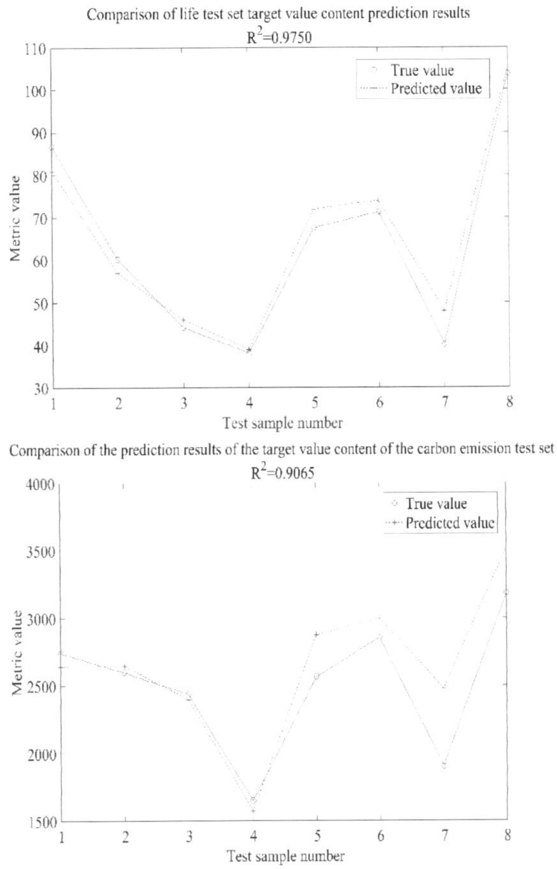

Figure 1. Comparison of life and carbon emission prediction results of BP neural network

4. Construction of multi-objective optimization model of NSGA-II algorithm based on BP neural network

NSGA-II (second-generation non-dominated ordering) is a genetic algorithm with good convergence and robustness for solving multi-objective optimization problems. Aiming at the optimization problem of CNC milling process parameters, different combinations of process parameters are regarded as an individual in the population. Each individual contains 3 process parameters: feed f, cutting depth a_p, cutting speed V_c. The objective function of the constructed NSGA-II main model is the BP neural network prediction model that predicts tool life T and process carbon emission C. Therefore, the optimization variable is the input parameter of the BP neural network (that is, the process parameter to be optimized), and the optimization target is the prediction target of the BP neural network. In summary, the mathematical model for the optimization problem of CNC milling process parameters is as follows:

$$F\{BP(f, a_p, V_c)\} = (maxT, min\,C) \tag{3}$$

$$\begin{cases} f_{min} \le f \le f_{max} \\ a_{pmin} \le a_p \le a_{pmax} \\ V_{cmin} \le V_c \le V_{cmax} \end{cases} \tag{4}$$

where F represents the nonlinear mapping relationship between the input and output of the NSGA-II optimization model; f, a_p, and V_c are the variables to be optimized for the multi-objective optimization model, and their input ranges are as shown in Equation (4); For the range of variables to be optimized, industrial applicability is used as a guide, i.e., 0.25 mm/rev<f<0.5 mm/rev, 0.75 mm<a_p<1.5 mm, 50 m/min<V_c<500 m/min. The specific steps are as follows:

(1) The optimization variables for solving the optimization problem of CNC milling process parameters are determined. By analyzing the optimization objectives, the process parameters that affect the optimization objectives are determined.

(2) A BP neural network prediction model is built. Taking the process parameters determined in step (1) as input, carbon emissions, and tool life as outputs, a neural network tool life and carbon emission prediction model are established. The prediction results of the established predictive model are trained, tested, and analyzed.

(3) The parameters of the NSGA-II multi-objective optimization host model are set, including the number of populations, the maximum number of iterations, the crossover probability, the mutation probability, etc.

(4) The population is initialized.

(5) The BP neural network is called the objective function of the NSGA-II multi-objective optimization main model.

(6) The NSGA-II algorithm process is entered until the end of the iteration.

(7) The process parameters in the last generation population and the corresponding optimization target value are obtained. The obtained process parameters are the Pareto optimal solution for the problem to be optimized.

The maximum number of iterations maxit=100, population size npop=50, crossover ratio pc=0.8, mutation probability mu=0.05, and the Pareto front optimal solution set is obtained after iteration to the maximum number of iterations.

The final set of Pareto front optimal solutions contains 19 parameters to determine the optimal solution. The optimal solution set obtained after optimization is determined by the Topsis method. The final Topsis optimal solution decision results are shown in Table 3.

Through the optimal solution decision results of the entropy weight Topsis method, it is concluded that the 5th group solution in the Pareto solution set is the optimal solution of the optimization problem of CNC milling process parameters, that is, a_p=1.039 mm, f=0.383 mm/rev, V_c=98 m/min. At this time, the process carbon emission C=1025.1421 $kgCO_2e$ and tool life T=120.6487 min are reduced by 61.17% compared with the optimal scheme a_p=0.75 mm, f=0.25 mm/rev, V_c=200 m/min. The process carbon emissions C=2640.0978 $kgCO_2e$ decreases by 61.17%, and the tool life T=81min increases by 48.95%. The optimization results show that the optimization scheme improves tool life and reduces the carbon footprint during machining.

Table 3. Topsis optimal solution decision results

Number	a_p/ (mm)	F/ (mm/rev)	V_c/ (m/min)	C/ ($kgCO_2e$)	T/ (min)	Optimal solution ordering
1	1.039	0.388	154	1166.4246	124.0863	2
2	1.081	0.375	352	1948.8819	137.3168	8
3	1.081	0.376	184	1872.2229	136.3569	7
4	1.012	0.375	352	2295.1342	139.7617	10
5	1.039	0.383	98	1025.1421	120.6487	1
6	1.039	0.376	106	2149.5446	138.7982	9
7	0.754	0.35	118	3403.637	147.298	18
8	1.081	0.376	184	1483.1665	130.7919	4
9	0.913	0.375	352	2788.3855	140.7734	13
10	1.147	0.375	352	1613.4833	132.2115	6
11	0.953	0.376	103	2579.0628	140.1445	12
12	0.754	0.35	335	3528.5683	149.1912	19
13	1.147	0.375	352	1613.4833	132.2115	5
14	0.754	0.375	190	3087.3275	143.5168	15

15	0.758	0.376	189	3067.4551	143.0409	14
16	0.754	0.376	189	3294.8861	144.1921	16
17	0.754	0.375	352	3303.6861	144.634	17
18	0.953	0.376	390	2579.0628	140.1445	11
19	1.012	0.375	352	1408.9461	127.5661	3

5. Conclusion

In this paper, a BP neural network prediction model is established using different process parameter combinations as input, CNC milling tool life and process carbon emissions as output. Through this method, the hidden relationship between process parameters and CNC milling tool life and process carbon emissions is found, and a predictive model is established. The coefficients of determination of the forecast targets are 0.97502 and 0.9065, respectively, enabling accurate predictions of tool life and process carbon emissions. Taking the maximum tool life and minimum process carbon emissions as the optimization goals, the NSGA-II multi-objective optimization main model of process parameters is constructed. Then, the constructed prediction network is called the objective function of the main model and optimized to solve. Finally, the Pareto front optimal solution set for the numerical control milling parameter optimization problem proposed in this paper is obtained. Using the Topsis method to decide on the optimal solution of the Pareto front optimal solution set, the optimal solution is finally obtained as a_p=1.039 mm, f=0.383 mm/rev, V_c=98 m/min. The result is an increase in tool life and a significant reduction in carbon emissions during machining.

References

[1] Rao R V Pawar P J. Parameter optimization of a multi-pass milling process using non-traditional optimization algorithms[J]. *Applied Soft Computing*, 2010, 10(2): 445-456.

[2] LIU Yifan, YAN Chunping, NI Hengxin, MOU Yun. Multi-objective optimization decision of high-speed dry cutting hobbing process parameters based on GABP and improved NSGA-II. [J]. *China Mechanical Engineering*,2021,32(09):1043-1050.)

[3] ZHANG Jiexiang, LI Zhiyong, ZHANG Wei, LIU Jianhou, SONG Shan, LI Yue. Optimization of side plate process parameters of micro-milled radial tire mold based on parallel GABP and NSGA-II[J]. *Manufacturing Technology and Machine Tool*,2021(07): 104-110.10.19287/j.cnki.1005-2402.2021.07.020.)

[4] CHEN Jinjiang, SUN Bo, JI Jinglan, JIAO Shilong, WANG Fang, TIAN Guangwei. Multi-objective optimization of high-speed milling parameters based on NSGA-II. algorithm[J]. *Machine Tool and Hydraulics*,2016,44(19):56-58+177.)

[5] Mativenga P T, Rajemi M F. Calculation of optimum cutting parameters based on minimum energy footprint[J]. *CIRP Annals - Manufacturing Technology*, 2011, 60(1):149-152.

[6] LI Congbo, FU Song, CHEN Administrative, JI Qianqian. Multi-objective optimization model of CNC gear hobbing parameters for high efficiency and energy saving[J]. *Computer Integrated Manufacturing Systems*,2020,26(03):676-687.)

[7] CHEN Jianlin, YAN Chunping, HOU Yuehui, CHEN Liang. Multi-objective high-speed dry cutting hobbing process parameter optimization decision-making method[J/OL]. *Mechanical Science and Technology*:1-11[2021-09-1409:54]. https: lldoi.org/10.13433/i.cnki.1003-8728.20200507

[8] YAN Kaiqiang, HUANG Xiaobin, ZHANG Shijie, SUN Baoyu, GAO Le. Simulation of the influence of cutting data on cutting force when milling titanium alloy by face milling cutter[J]. *Tool Technology*,2022,56(04):80-83.)

[9] A. Agogino, K. Goebel, Milling Data Set, NASA Ames Progn. Data Repos. (2007). http:/0t.arc.nasa.gowprojectiprognostic-data-repository.

[10] Bao Ziyang, Yu Jizhou, Yang Shan. Intelligent optimization algorithm and its MATLAB example (2nd edition) [M]. Beijing: *Publishing House of Electronics Industry*,2018:185.)

The Grinding Parameter Optimization under the Carbon Benefit Objective

Yingquan Tian, Ruixue Yin

School of Mechanical Engineering, Guizhou University, Guizhou Guiyang 550025

e-mail:996502448@qq.com e-mail:yinruixue@sina.com

Abstract. Grinding is a high carbon emission process in processing technology. To improve the energy efficiency of the grinding process and reduce the carbon emissions and costs of the grinding process, the concept of carbon efficiency is proposed. The study takes the maximisation of carbon benefit in the grinding process as the optimization objective. The linear speed of the grinding wheel, the feed speed of the workpiece and the radial feed speed of the grinding wheel are taken as the optimization variables. Considering the grinding performance of the grinding wheel, taking the performance of the machine tool and the machining requirements of the workpiece as constraints, a multi-objective optimization model of the grinding process is established. The model is optimized by a genetic algorithm, and the effectiveness of this method is verified by a case.

1. Introduction

To cope with climate change and future unfair carbon taxes and carbon tariffs, the Chinese government has proposed the '30·60' dual carbon target [1]. Carbon emissions research in all walks of life has become a hot spot. As China's main energy consumption and carbon emissions industry, the manufacturing industry accounts for 57% of energy consumption and carbon emissions [2]. The research on carbon emissions of the manufacturing industry will be of great significance in achieving the dual carbon target.

In the previous low-carbon optimization process of grinding parameters, the carbon emissions related to machine tool processing are mainly considered. In contrast, the carbon emissions generated using grinding wheels are relatively insufficient [3]. To obtain more accurate estimation results of carbon emissions in the grinding process, it is necessary to study the carbon emissions of the grinding wheel dressing process [4].

2. Carbon emission evaluation model of the grinding process

The grinding process is a typical process of machining [5]. Carbon emissions mainly include carbon emissions caused by the power consumption of grinders and carbon emissions caused by loss of cutting fluid and grinding wheel [6]. The proportion of the rest is very small and difficult to quantify, which can be ignored. The calculation method of total carbon emission in the grinding process is as follows:

$$C=C_e+C_m+C_s \tag{1}$$

where C_e is carbon emissions caused by grinding machine power consumption (gco2e); C_m is carbon emissions from grinding fluid consumption (gco2e); C_s is carbon emissions caused by wheel wear (gco2e).

2.1 Estimation of carbon emissions from grinding energy

Grinder from start to work is generally divided into four stages, start-up stage, standby stage, no-load stage, and processing stage. According to its characteristics, electric energy carbon emissions can be divided into constant carbon emissions. C_{ec} (gco2e) and variable carbon emissions C_{eb} (gco2e) are expressed as Equation (2).

$$C_e = C_{ec} + C_{eb} \tag{2}$$

Because the time of machine start-up, standby stage and no-load stage cannot be optimized by cutting parameters, the influence of these stages is not considered. Grinding energy carbon emissions are expressed as follows:

$$C_e = f_{ele} \sum_{i=1}^{m} P_i t_i + f_{ele} \int_0^{t_{eq}} (P_{u0} + m_1 n + m_2 n^2) dt + f_{ele}(1 + k_{ef}) \int_0^{t_{eq}} a_1 \cdot v_w^{b1} \cdot f^{c1} \cdot v_s^{d1} \cdot v_s dt \tag{3}$$

where f_{ele} is the carbon emission factor of electric energy ($gco2e/kW \cdot h$), i is the i-th constant system, m means there are m constant systems, P_i is the power of a constant system (kw), t_{ek} is the no-load time (h), P_{u0} is the minimum no-load power (kw), m_1 and m_2 are the spindle speed coefficients, t_{eq} is the grinding time (h), $a1$、$b1$、$c1$、$d1$ are the grinding force correlation coefficient, k_{ef} is the additional loss coefficient, t_i is the operating time of a constant system (h).

2.2 Estimation of carbon emission of grinding fluid

Grinding fluid is essential when grinding a workpiece. The carbon emission of grinding fluid is calculated as Equation (4) [7]:

$$C_m = \frac{t_{eq}}{T_m}(f_{oil} + f_{wc}) \times Q \tag{4}$$

where T_m is the grinding fluid replacement cycle used (h). According to the calculation of one month, 14 hours a day, T_m=320h; f_{oil} is the carbon emissions from grinding fluid production ($gco2e/L$); f_{wc} is the carbon emissions from grinding fluid treatment ($gco2e/L$); Q is the initial amount of cutting fluid, determined by the capacity of the grinding tank, 500 L.

2.3 Estimation of carbon emission of grinding wheel loss

The loss of the grinding wheel during the grinding process is mainly the wear loss of the grinding wheel during the grinding process and the volume loss of the grinding wheel due to dressing [8].
The total carbon emission of grinding wheel wear during the grinding of a single part can be expressed as Equation (5):

$$C_s = C_{ss} + C_{sx} = f_s V_s + (C_{sxe} + C_{sxs}) / n_D \tag{5}$$

where f_s is the carbon emission factor of the grinding wheel ($gco2e/mm^3$); V_s is the wear loss volume of the grinding wheel in the grinding process (mm^3); C_{ss} is the carbon emission of grinding wheel wear loss in grinding process; (gco2e) C_{sx} is the total carbon emissions of grinding wheel dressing process (gco2e); n_D is the number of grinding parts after each dressing; C_{sxe} is the grinding wheel dressing energy carbon emissions (gco2e); C_{sxs} is the grinding wheel dressing loss carbon emissions; (gco2e) n_D is the number of grindable parts after one wheel dressing

The carbon emission of grinding wheel wear loss during grinding is as follows.

$$C_{ss} = f_s \frac{V_g}{G_0(\frac{Q_w'}{v_s})^{-g}} = f_s \frac{V_g}{G_0(\frac{\pi d_w f}{v_s})^{-g}} \tag{6}$$

where d_w is the workpiece diameter (mm), G_0 g is the empirical constant, and Q_w' is the material removal rate per unit width (mm^3/s).

When the grinding wheel grinds the workpiece, it must be dressed first. The carbon emission of wheel dressing is as follows.

$$C_{sx} = (f_{ele}P_{sxe}t_{sxe} + f_s\pi Ba_d d_s)/ \\ [(n_0-1)\ln(\frac{R_{a,0}-R_g(\pi d_w f)^j}{R_{a\max}-R_g(\pi d_w f)^j})+1] \tag{7}$$

where P_{sxe} is the grinding wheel dressing power (kW); t_{sxe} is the grinding wheel dressing time (h); B is the grinding wheel width (mm); a_d is the grinding wheel dressing depth (mm); d_s the grinding wheel diameter (mm); n_0 is the change rate of surface roughness, which is constant; $R_{a\max}$ is the surface roughness required for parts (μm); $R_{a,0}$ is the initial roughness value of the machined part (μm) after the grinding wheel is dressed, $R_{a,0} \le R_{a\max}$; $R_{a,\infty}$ is the steady-state roughness value (μm); R_g is the correlation coefficient; j is the correlation coefficient, generally taken as 0.5.

In summary, considering the maximum number of grindable parts after grinding wheel dressing, the total carbon emission model of the grinding process can be expressed by Equation (8).

$$C = f_{ele}\sum_{i=1}^{m} P_i t_i + f_{ele}\int_0^{t_{eq}} (P_{u0} + m_1 n + m_2 n^2)dt + \\ f_{ele}(1+k_{ef})\int_0^{t_{eq}} a_1 \bullet v_w^{b1} \bullet f^{c1} \bullet v_s^{d1} \bullet v_s dt + \frac{t_{eq}}{T_m}(f_{oil} + \\ f_{wc}) \times Q + f_s \frac{V_g}{G_0(\frac{\pi d_w f}{v_s})^{-g}} + (f_{ele}P_{sxe}t_{sxe} + \\ f_s\pi Ba_d d_s)/[(n_0-1)\ln(\frac{R_{a,0}-R_g(\pi d_w f)^j}{R_{a\max}-R_g(\pi d_w f)^j})+1] \tag{8}$$

3. Multi-objective optimization model of grinding parameters based on carbon efficiency

3.1 Grinding cost model

The cost of grinding a single part mainly includes the cost of energy (M_e), labor cost (M_l), grinding fluid cost (M_m), grinding wheel cost (M_s), equipment cost (M_b) and carbon emission cost (M_C), other costs are not considered. The cost of grinding a single part can be expressed as Equation (9)

$$M = M_e + M_l + M_m + M_s + M_b + M_C \tag{9}$$

The energy cost of grinding is the cost of electricity consumption in the grinding process, which can be expressed as Equation (10):

$$M_e = E_e k_e \tag{10}$$

where k_e is the unit price of electricity. Based on industrial peak hour electricity charges, it is 1.025 yuan/kw.

Labor cost refers to the salary paid to workers. Labor cost can be expressed as the product of workers salary per unit time and processing time. The calculation is as follows:

$$M_r = k_r t_{eq} \tag{11}$$

where k_r is worker's salary per unit time (/h), which is calculated at 20/h for simplified calculation.

The cost of grinding fluid mainly includes the preparation cost of grinding fluid and the processing cost of grinding fluid. The cost of grinding fluid can be expressed as Equation (12).

$$M_m = \frac{t_{eq}}{T_c}(k_m + k_h)Q \tag{12}$$

where k_m is the unit price of grinding fluid after dilution $(yuan/L)$; k_h is the unit price of grinding fluid treatment $(yuan/L)$.

The grinding wheel cost includes the grinding wheel wear cost M_{sq}, the grinding wheel dressing power cost M_{sxe}, the grinding wheel dressing loss cost M_{sxs}, and the grinding wheel dressing labor cost M_{sxr}, which are calculated by Equation (13).

$$M_s = k_s \frac{V_g}{G_0(\frac{\pi d_w f}{v_s})^{-g}} + (M_{sxe} + M_{sxx} + M_{sxr}) / n_D \tag{13}$$

where k_s is the unit volume grinding wheel price $(yuan/mm^3)$. The ratio of the unit price of the grinding wheel to the total available volume of the grinding wheel is calculated.

The use cost of grinding equipment is mainly the use cost of the grinding machine. The cost of the grinding machine is converted to the grinding process according to the time standard. Because the recovery cost of the machine tool is different from the treatment method, it is not considered. The cost of using the equipment is calculated as follows.

$$M_b = \frac{t_{eq}}{T_b} k_b \tag{14}$$

where k_b is the quipment purchase cost (yuan); T_b is the machine tool life (h).

Taking carbon emission right as a commodity forms the trading of carbon emission right [9]. Reducing carbon emissions in the production process can save some costs for enterprises. Grinding carbon emission cost can be expressed as Equation (15):

$$M_C = k_C C \tag{15}$$

where k_C is the carbon tax price $(yuan/gco2e)$.

For a specific part, the cost of measuring tool and fixture wear in actual machining is very small compared with the total cost of machining in the grinding process, which can be ignored. The material cost depends on the mining, transportation and production of the material, and has nothing to do with the cutting parameters. Therefore, it is not considered.

3.2 Carbon benefit

With reference to Yin et al.'s study [10], the carbon benefit is defined as the benefit value obtained by the unit carbon emission generated during the processing process, such as Equation (16):

$$CB = \frac{K - M}{C} \tag{16}$$

where CB is the carbon benefit of grinding $(yuan/gco2e)$; K is the contribution value of the process to the price of the processed parts.

It can be seen from the definition that the greater the carbon benefit is, the greater the benefit obtained in the case of unit carbon emissions.

3.3 Multi-objective optimization model of grinding parameters based on carbon efficiency

The problem of maximizing carbon efficiency can be transformed into the problem of minimizing carbon emissions and grinding costs in the grinding process. The objective function is in Equation (17):

$$\min F(v_s, f, v_w) = (\min C, \min M)^T \tag{17}$$

The optimization of carbon efficiency is essentially a multi-objective optimization problem. The commonly used optimization method to transform the multi-objective problem into a single objective is the weight coefficient method. The transformed single objective function is in Equation (18):

$$\min F(v_s, f, v_w) = (\omega_1 \min C + \omega_2 \min M)^T \tag{18}$$

where ω_1、ω_2 are the weight coefficients, and $\omega_1 + \omega_2 = 1$. Generally, the two are regarded as equally important, and their values are 0.5.

Unified dimension processings are in Equations (19) (20):

$$C^* = \frac{C - C_{\min}}{C_{\max} - C_{\min}} \tag{19}$$

$$M^* = \frac{M - M_{\min}}{M_{\max} - M_{\min}} \tag{20}$$

where C^* is the carbon emissions after unified dimension; M^* is the cost after unifying dimension; $C_{\min}$ and $C_{\max}$ are the minimum and maximum values of carbon emissions; $M_{\min}$ and $M_{\max}$ are the minimum and maximum cost.

The objective function after dimension unification is as follows.

$$\min F(v_s, f, v_w) = (\omega_1 \min C^* + \omega_2 \min M^*)^T \tag{21}$$

4. Case

4.1 Model parameter setting and grinding experiment

The experiment uses MORARA MH 3000 D CNC grinding machine to grind a certain type of shaft parts. The experimental results are shown in Table 1.The error between the theoretical value and the measured value of carbon emission is within 5%. It can be seen that although the carbon emission model of NC grinding established above has some errors, the carbon emission model of NC grinding established in this paper is effective considering the complexity of actual processing and measurement errors of experimental data. It has certain guiding significance for carbon emission in actual processing.

Table 1.Experimental results

Number	Grinding wheel speed (m/s)	Radial feed f (mm/min)	Workpiece rotation speed a_e (r/mm)	Grinding wheel motor current, I (A) Theoretical value/actual value	Carbon emission (kg) Theoretical value/actual value
1	35	0.01	15	5.79/5.92	23.43/24.41
2	40	0.01	20	7.28/54.8	23.68/23.74
3	45	0.01	30	9.80/9.46	24.07/21.6
4	30	0.05	15	18.30/18.75	6.09/6.44
5	40	0.05	30	31.47/33.21	6.52/6.37
6	45	0.05	20	25.55/26.14	6.33/6.27
7	30	0.10	15	30.82/31.95	3.88/3.93

8	35	0.10	20	39.11/40.03	4.01/4.03
9	45	0.10	30	55.42/53.48	4.27/4.30
10	30	0.15	15	41.81/40.97	3.12/3.22
11	35	0.15	20	53.05/54.28	3.24/3.21
12	40	0.15	30	71.88/72.64	3.44/3.54

4.2 Optimization algorithm and result analysis

In order to verify the practicability of the established carbon emission and processing cost model, the grinding process parameters are optimized.

The NSGA-II algorithm is used to optimize the model. The number of variables in the optimization experiment is 3, the number of optimization targets is 2, the crossover ratio of the algorithm is set to 0.8, the number of populations is 50, the maximum number of iterations is 200, and the mutation probability is 0.05.

The Pareto solution set is not relatively optimal and needs to be weighted. In the optimization process. When the carbon emission weight is 0.5 and the cost weight is 0.5, the optimal cutting parameters combination are as follows: the grinding wheel linear speed is 30 m/s, the radial feed speed is 0.11 mm/min, and the workpiece rotation feed speed is 15r/min. At this time, the carbon emission is 3052.21 g, and the cost is 37.93 yuan. Compared with the empirical method, carbon emissions decreased by 16.44 %, and processing costs decreased by 9.11%. Therefore, the optimized cutting parameters greatly utilize the processing capacity of the CNC machine tool, reducing the processing carbon emissions and saving the cost.

References

[1] Yin Wei-hua. The impact of carbon tax collection and policy choice under different emission reduction policies: Based on the analysis of carbon peak and carbon neutral goals [J/OL]. *Journal of Guangdong University of Finance and Economics*,2021(05):16-26.

[2] National Bureau of Statistics of the People's Republic of China. China Statistical Yearbook 2014[M]. *Beijing：China Statistics Press*,2014.

[3] Marius Winter, Wen Li, Sami Kara,et al. Determining optimal process parameters to increase the eco-efficiency of grinding processes[J]. *Journal of Cleaner Production*, 2014,66:644-654.

[4] Jeffrey Badger, Stuart Murphy, Garret E. O'Donnell. Big-and-dull or small-and-sharp: A comparison of specific energy, wheel wear, surface-generation mechanisms and surface characteristics when grinding with Al 2 O 3 and CBN to achieve a given surface roughness[J]. *Journal of Materials Processing Tech.*,2021,288.

[5] Liu Wei, Shang Yuanyuan, Deng Zhaohui, Liu Rentong, Xiao Zhouqiang. Parameter Optimization of High-Speed Cylindrical Grinding of Bearing Steel GCr15 Based on Response Surface Method [J]. *China Mechanical Engineering*,2019,30(23):2829-2834.

[6] Guo Deng-yue. Quantitative Analysis and Optimization of Energy Consumption And Carbon Footprint In Grinding Processes[D]. *Harbin Institute of Technology*,2013.

[7] Hu Shao-hua, Liu Fei, He Yan, Hu Tong. Study on the Characteristics ofNo-load Energy Parameters of Variable Frequency Main Drive System ofCNC Machine Tools[J]. *Computer Integrated Manufacturing Systems*,2012,18(2):326-331.

[8] Liu Fei, Xu Zong-jun, Dan Bin. Energy Characteristics of Mechanical Processing System and Its Application[M]. *Beijing: Mechanical IndustryPress*,1995:8-39.

[9] Stephen Malkin, Cai Guang-qi, Gong Ya-dong,Song Gui-liang. Grinding Technology Theory and application of machining with abrasives[M].*Shen Yang: Northeastern University Press*,2002：137-216.

[10] Yin Rui-xue, Wu Wei, Chen Wen-qing. Study on cutting parameters optimization model for NC turning process based on carbon benefit [J]. *Modern manufacturing Engineering*,2018(10):17-22.

Journal of Physics: Conference Series **2561** (2023) 012017 doi:10.1088/1742-6596/2561/1/012017

Fault diagnosis of train control onboard equipment based on Attention-Capsule Network

Xin Zhao[1,a], **Yunshui Zheng**[2]

[1,2] School of Automation and Electrical Engineering, Lanzhou Jiaotong University, CHN

[a]E-mile:1536883772@qq.com

Abstract. The onboard equipment of the train is a key component to ensure the safe and efficient operation of the train. Therefore, improving the fault detection ability of onboard equipment is crucial for the safe operation of trains. The driving log record file (AE-LOG) was generated by the CTCS-300T series control system (AE-LOG) as the starting point, and the improved capsule network (CAPSNET) as the research method was studied. By using the capsule network, this model greatly improves the attributes of text characteristics in high-dimensional space and low -dimensional space. At the same time, the routing method of the fusion attention mechanism is used to improve the efficiency of feature selection. Through experimental verification of operating data provided by a certain road bureau, and comparing with traditional models, the accuracy of the model in the control vehicle device fault diagnosis of the control of the vehicle is significantly improved, which shows the effectiveness of the model.

1. Introduction

Railway signaling equipment is an important guarantee for ensuring the safety of train operation [1]. The train control system shoulders an important responsibility for train safety. Early research on faults in onboard equipment mostly relied on the work experience of on-site staff for classification, resulting in inaccurate classification and low efficiency.

More and more scholars have proposed to use intelligent technology to achieve the intelligence of railway systems [2-4]. Chen Guangwu [5] and others proposed a method based on RS-ICS-BP, using rough set theory to eliminate the Rong redundancy condition and avoid local minimum problems through ICS optimization. Yang Lianbao [6] and others conducted fault research on railway signaling equipment based on unbalanced text, taking into account the characteristics of different classifiers and conducting integrated learning through voting. Zhao Qinlu [7] and others have effectively improved the accuracy of the experiment by using a layered attention mechanism network layer to select text words and sentences that rely on a bidirectional long and short memory network (LSTM) architecture. Most of the above scholars have used traditional neural networks for fault diagnosis, resulting in sparse text semantic features and ineffective utilization of spatial attributes between low-dimensional and high-dimensional features.

In previous studies, Fault diagnosis methods based on CNN (Convolutional Neural Network) are widely used in text-based fault diagnosis [8-10], and various variants based on CNN emerge in an endless stream. However, CNN is not perfect. During pooling operations, some low-level features will be lost. Based on this issue, Professor Hinton proposed the capsule network[11]. Experiments have shown that the performance of the capsule network is much higher than that of LSTM, CNN, etc.

This article uses capsule networks as the basic architecture for fault diagnosis and intelligent classification. Through cleaning on-site data, the text to be processed is obtained. The processed corpus is input into the currently popular pre-trained word vector model to obtain the word vector through training. The word vector forms a text matrix in the order of its position in the text and is input into the diagnostic model for diagnosis and classification. In the diagnostic model, the attention mechanism is combined with the routing mechanism to enable sub-capsules to find more interesting parent capsules by iterating the weights of different parent capsules. By comparing with existing fault diagnosis models for onboard equipment, the superiority of the model is reflected in this article.

2. Research object

This paper takes the CTCS-300T vehicle subsystem as an example for research. The structural composition of CTCS-300T is shown in Figure 1.

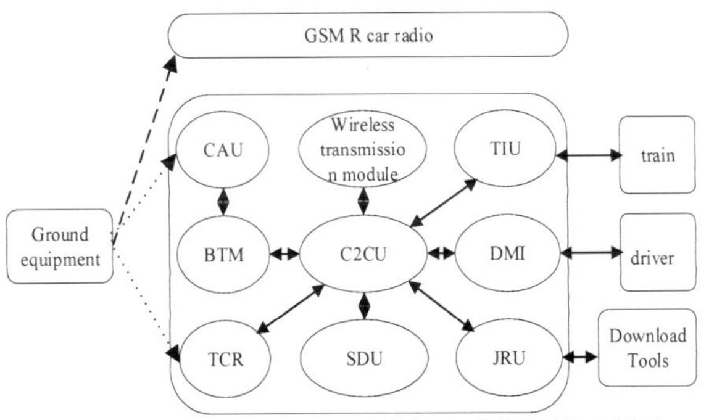

Figure 1. Structural composition of CTCS3-300T

This article takes the AE log text generated by the train control onboard device during train operation as the dataset and takes the balize information receiving unit (BTM) module as the research object. The first line of the AE log text contains the text sequence number, file name, task number, and so on. The second line contains the fault code and the recorded text of the currently running event.

3. Diagnosis Model Based on Capsule Network

Capsule Network was initially used in image processing and was first applied to text mining research in 2018. This article takes the capsule network as the core framework and builds a diagnostic model as shown in Figure 2.

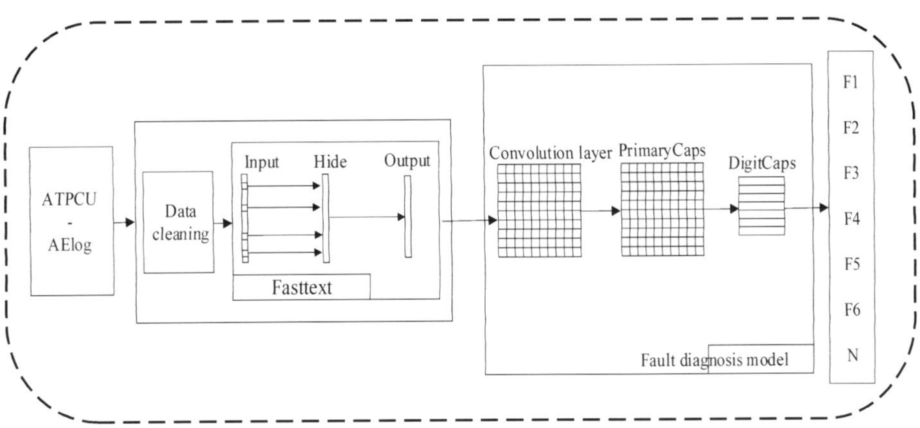

Figure 2. Diagnostic Model Structure Diagram.

3.1. Data handling

The data processing process is divided into two parts. First, it is to clean the original text in the dataset. This article uses the regularization method to filter out irrelevant noise such as the text sequence number, date, and time generated by the system in the AE-log text, and perform word segmentation to generate a thesaurus and arrange it based on frequency. Some of the thesauri are shown in Figure 3.

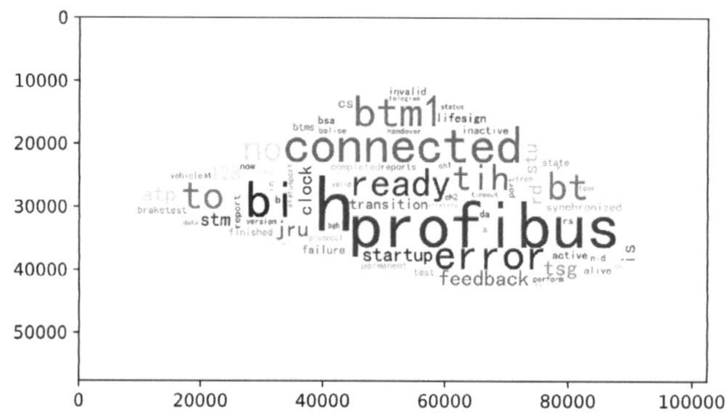

Figure 3. Some high-frequency words in the thesaurus.

This article uses the currently popular pre-processing corpus to generate word vectors. Two kinds of pre-trained word vectors, Glove, and Fast text, are used to load word vectors. By detecting the coverage of these two types of pre-trained word vectors on the experimental dataset, as shown in Table 1, under the four conditions of unprocessed data, lowercase text characters, remove special characters, and remove abbreviations, the overall performance of fast text is more stable. Therefore, this article uses the method of fast text to embed word vectors

Table 1. Pre-training word vector coverage.

Pretraining word vector	Unprocessed data		All text characters lowercase		Remove special characters	
	Word coverage	Text coverage	Word coverage	Text coverage	Word coverage	Text coverage
glove-	40.34%	43.15%	74.77%	76.41%	89.91%	91.29%
fast text	73.31%	77.05%	72.01%	67.85%	87.16%	88.71%

3.2. Diagnostic model

This diagnostic model includes three layers, namely, the local feature extraction layer, data type conversion layer, and digital capsule layer.

The local feature extraction layer of the first layer is a conventional convolution layer, consisting of multiple convolution kernels of size and step size of 1 and extracting multiple low-level features at different locations in the input text matrix through multiple convolution kernels of different sizes.

The data type conversion layer receives shallow features from convolutional layers, expands the received features, and merges their corresponding positions, outputting them as capsule vectors, completing the transition from quantity input and output to vector input and output.

In the Digit Caps, six digital capsules are set for the fault types of BTM devices. Each capsule represents a fault category for the BTM device.

The essence of the attention mechanism is to focus attention on a few important pieces of information and change the focus degree of its corresponding value through the change of weight. In the model of this topic, the attention mechanism and routing mechanism are combined to give weight to lower-level capsules, to achieve an autonomous selection of sub-capsules versus parent capsules, as shown in Figure 4.

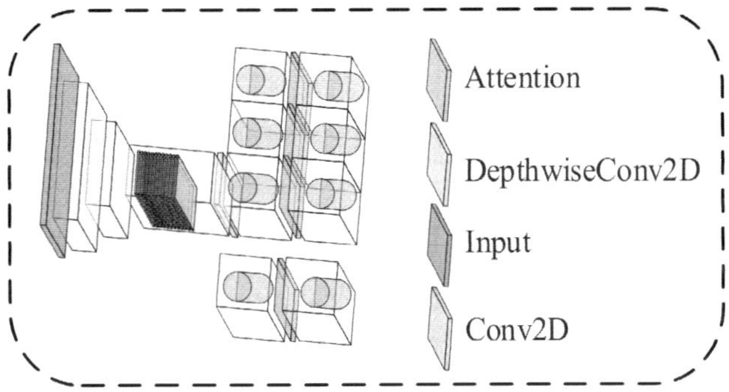

Figure 4 Routing Method Integrating Attention Mechanism.

4. Simulation experiment and performance evaluation

4.1. Data source and experimental platform

The data set in this article is derived from part of the actual field data provided by a railway bureau from 2019 to 2021, with a total of 672 sets of fault texts. Randomly select 75% of each category in the dataset by sample category as a training set and 25% as a test set.

4.2. Data source and experimental platform

The word vector dimension used in the experimental part of this topic is 300, the convolutional layer convolution kernel is $2\times300, 3\times300, 4\times300, 5\times300$, the number of convolutional cores in each group is 100, the number of epochs is 100, and the batch size is 25. Figure 5 shows the trend of changes in various performance indicators under different epochs.

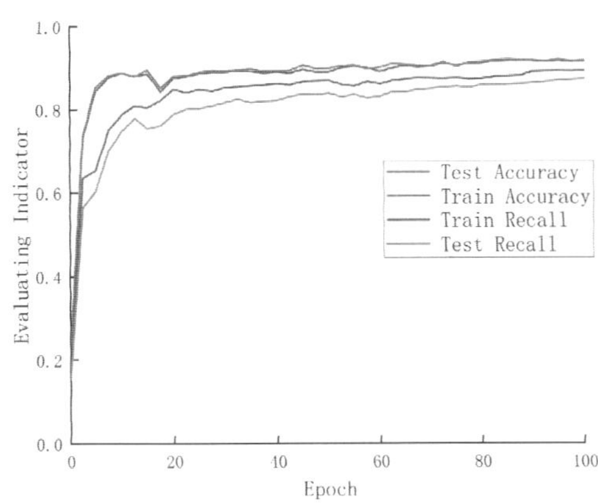

Figure 5. Performance index changes under different iterations.

4.3. Performance Verification

The performance of this model and other models on this dataset was compared. It mainly includes CNN, SVM, KNN, RF, and other models. Table 2 shows the effectiveness of these models.

Table 2. Performance Comparison.

Model	Precision	Recall	F-Measure
CNN	0.7883	0.7198	0.7525
LSTM	0.8677	0.7962	0.8304
Bi-LSTM	0.8463	0.8249	0.8355
SVM	0.8551	0.6689	0.7506
Attention-Capsnet	0.9041	0.8795	0.8916

According to the above table, the improved capsule network performs well in all three performance indicators, with a precision rate of 90.41%

5. Conclusions

(1) This paper introduces a diagnosis model based on an improved capsule network, with AE log text as the data set. The structural transformation of AE log data from text to the matrix is achieved through data cleaning of AE log text and embedding of dataset word vectors using the existing pre-training word vector tool Fast Text to generate a text matrix. Extracting rich local features from text data at different positions in a text matrix uses convolutional kernels of different widths. The sub-capsule weights are trained and learned through a combination of routing and attention mechanisms. And the output in the subsequent diagnostic model is completed.

(2) Through experiments using real driving data on-site, using three performance indicators, Precision, Recall, and F-Measure, to evaluate the model, and display based on the results, the improved capsule network has a good effect in the field of AE-log text-based vehicle equipment fault diagnosis, with a precision rate of 90.41%.

References

[1] TONG, L. B., (2016) "Introduction of Railway," China Railway Publishing House Co., Ltd, Beijing. Https://ss.zhizhen.com/s?sw=%E9%93%81%E9%81%93%E6%A6%82%E8%AE%BA%E7%AC%AC7%E7%89%88&size=15&isort=0&x=0_351

[2] WANG, T. J., (2018) "Overall Architecture and Development Prospect of Intelligent Railway," SPB Publishing, 27(07):1-8. DOI: 10.396 9/j. issn.1005-8451.2018.07.003.

[3] XIONG, J. Y., SHEN, Z. Y., (2021) "Rise and future development of Chinese high-speed railway," Journal of Transportation Engineering, 21(05):6-29. Https://kns.cnki.net/kcms2/article/abstract?v= 3uoqIhG8C44YLTlOAiTRKibYlV5Vjs7iy_Rpms2pqwbFRRUtoUImHballSiW2K0mMYwiydVgfCRjH8GR6M89S6GjVo1LWufh&uniplatform=NZKPT.

[4] NING, B., MO, Z.S., LI, K.C., (2019) "Application and Development of Intelligent Technology for High-Speed Railway Signal System," Journal of the China Railway Society, 41(03):1-9. Https://kns.cnki.net/kcms2/article/abstract?v=3uoqIhG8C44YLTlOAiTRKibYlV5Vjs7iLik5jEcCI09uHa3oBxtWoFi74vRfvwtQmlGPU9P0N7eKPEpyITtyFEQv8KwWF73S&uniplatform=NZKPT

[5] NIU, H.X., GUO, Z.Y., CHEN, G.W., (2020) "Fault diagnosis of onboard BTM based on RS-ICS-BP," Journal of Beijing Jiaotong University, 44(02):52-57, Https://kns. cnki.net/kcms2/article/abstract?v=3uoqIhG8C44YLTlOAiTRKibYlV5Vjs7i8oRR1PAr7RxjuAJk4dHXojxZ6ZbKllCeBzKcQaILk0je_qFhiXyJZfXJsdrSr-8N&uniplatform=NZKPT

[6] YANG, L. B., LI, P., XUE, R., MA, X.L., WU, Y.H., ZOU, D., (2018) "Intelligent Classification of Railway Signal Equipment Based on Imbalances Text Data Mining," Journal of the China Railway Society, 40(02):59-66. Https://kns.cnki.net/kcms2/article/abstract?v=3uoqIhG8C44YLTl OAiTRKibYlV5Vjs7i0-kJR0HYBJ80QN9L51zrP7UlXq79m8uiiGGdbzTwQEsGeLr4VTkxFPJtR1aQZ6mi&uniplatform=NZKPT.

[7] ZHAO, Q. L., CAI, X. D., LI, B., LV, L., (2018) "Text feature extraction method based on LSTM-Attention neural network," Modern Electronics Technique, 41(08):167-170.

Https://kns.cnki.net/ kcms2/article/abstract?v=3uoqIhG8C44YLTlOAiTRKibYlV5Vjs7i0-kJR0HYBJ80QN9L51zrP2 IALhzxzd-_hqetw3eog1_hL_6bvUX_w3QvdghXXD3U&uniplatform=NZKPT

[8] LU, R. J., LIN, H. X., XU, L., LU, R., ZHAO, Z. X., BAI, W.S., (2022) "Fault diagnosis for onboard equipment of train control system based on CNN and PSO-SVM hybrid model," Journal of Measurement Science and Instrumentation. 13(04):430-438. Https://kns.cnki.net/kcms2/article/abstract?v=3uoqIhG8C44YLTlOAiTRKibYlV5Vjs7iJTKGjg9uTdeTsOI_ra5_XQI2llpms5oEfjmUVZaI51pKISzvaLH8dsXIr_5FnGeC&uniplatform=NZKPT

[9] ZHOU, L.J., DANG, J.W., WANG, Y.X., ZHANG, Z.H., (2020) "Fault Diagnosis for Onboard Equipment of Train Control System Based on CNN-CSRF Hybrid Model," Journal of the China Railway Society. 42(11):94-101. Https://kns.cnki.net/kcms2/article/abstract?v=3uoqIhG8C44YLT lOAiTRKibYlV5Vjs7i8oRR1PAr7RxjuAJk4dHXohRrbJzKeYe02jksYgTtl734CFRac1udiT-DsPY498wN&uniplatform=NZKPT

[10] LU, R.J., (2022) "Research on Combined Diagnosis Model of CNN and PSO-SVM for Onboard Equipment of Train Control System," Lanzhou Jiaotong University. DOI:10.27205/d. CNKI. Altec. 2022.001077.

[11] Hinton, G.E., Krizhevsky, A., Wang, S.D., (2011) "Transforming Auto-encoders," in International Conference on Artificial Neural Networks. Berlin: Springer. 44-51.

Long-distance intelligent liquid supply for coal mining faces based on liquid demand prediction

Ran Li [1,2], Wenshu Wei [2], Yuehua Lai [1,2]*, Dalong Wang [3], Bo Liu [2], Tongcheng Wang [2]

[1] Graduate School, China Coal Research Institute, Beijing, 100013, China

[2] Beijing Tianma Intelligent Control Technology Co., Ltd., Beijing, 100013, China

[3] School of Mechanical Electronic & Information Engineering, China University of Mining and Technology Beijing, Beijing, 100083, China

*Correspondence: laiyh@tdmarco.com

Abstract: The coal mine pump station liquid supply system adapts to changes in the main pressure for supply, but under conditions of long-distance liquid supply beyond 2000 m, there is a problem with the difficult system response to the hydraulic support liquid demand, resulting in significant pressure fluctuations in the hydraulic system and serious performance degradation of the supply system. To solve this problem, a long-distance intelligent liquid supply method based on liquid demand prediction for coal mining faces is proposed to ensure stable liquid supply. The method integrates parameters such as the mining machine's position, the hydraulic support's column pressure before and after the mining machine, the forward travel distance, and action codes to construct a feature vector. Based on the Transformer Model, a Liquid Demand Prediction Model for hydraulic support is established. Attention mechanisms are combined to solve the problem of time prediction for long sequences. The model is validated using real measured data from a hydraulic system in a coal mine, and the experimental results show that the model can accurately predict liquid demand and achieve long-distance intelligent liquid supply for coal mining faced by combining pump station control strategies.

1. Introduction

The pumping station supply system is a complete set of hydraulic power equipment that provides hydraulic power for the hydraulic supports' action and the coal mining machine's spray dust reduction as the power equipment for the mining face. It is the core part of the entire hydraulic system of the fully mechanized mining face [1, 2]. To solve the problems of maintenance intensity and safety in limited mining space, the development of long-distance intelligent supply technology based on the concept of "space decoupling" has been rapidly developed. It also puts higher requirements on long-distance intelligent control technology [3]. In recent years, researchers have carried out research on long-distance intelligent control. When supplying at a short distance, the pressure change of the system during the support action characterizes whether the current supply flow rate is sufficient and stable. Therefore, the pump station supply system adapts the supply based on the total pipeline pressure change [4, 5]. However, due to the significant increase in supply distance, the total pipeline pressure monitoring point is far from the hydraulic support. Using the total pipeline pressure to characterize the hydraulic support system pressure will cause the system's supply to respond slowly to the hydraulic support's liquid demand, resulting in insufficient and excessive alternating supply flow rates. Insufficient supply flow will cause pressure drop, slow support action, and even make it impossible to complete the scheduled action under load, affecting coal mining. Excessive supply flow will cause pressure fluctuations, causing support action tremors, affecting support control accuracy, and causing hydraulic technology problems such as overflow loss. Only by predicting hydraulic support action and controlling and adapting the supply flow rate in advance can the supply control dynamically adapt to the changing group actions of the hydraulic supports and realize intelligent long-distance supply for the mining face.

Therefore, this paper proposes a long-distance intelligent liquid supply method based on liquid demand prediction for coal mining. The traditional steady-state control mode based on main pipe pressure is changed by accurately predicting the demand for liquid and combining it with pump station control strategies. It has certain reference significance for the intelligent upgrading coal mine liquid supply systems.

2. Related works

Existing time series prediction methods can be roughly divided into traditional time series models and encoder-decoder prediction paradigms based on RNN and its variants. The encoder-decoder prediction paradigm based on RNN and its variants has achieved breakthrough performance. Since a single hydraulic support action includes column lifting, lowering, strut pulling, balancing, side protection, and bottom lifting, etc., different hydraulic supports also need to perform various grouped actions to follow the movement of the shearer. Therefore, when predicting the hydraulic support liquid demand, a sequence of actions before the hydraulic support needs to be considered. LSTM is a commonly used time series prediction model, and various improved variants have been widely used in predicting power loads, equipment life, and commodity prices [6, 7]. Studies have shown that the Transformer Model has parallel computing advantages and exhibits better long-term memory capabilities than LSTM in long-time series model prediction. The Transformer was used to predict the prices and sales volumes of various commodities on e-commerce platforms, which well reflects the Transformer's ability to handle long-term sequential information [8]. The Transformer Model was applied to pedestrian trajectory prediction and achieved better prediction results than LSTM [9]. In summary, the Transformer Model performs better than the RNN model in capturing long-term dependency relationships and is more suitable for predicting the hydraulic support liquid demand.

3. Preliminaries

Firstly, the liquid demand prediction of hydraulic support is defined. Rolling prediction of liquid demand is performed with a fixed window size, and the input at time t is defined as $\boldsymbol{x}^t = \left\{x_1^t, \cdots, x_{L_x}^t \middle| x_i^t \in \mathbb{R}^{d_x}\right\}$, and the corresponding sequence for the predicted results is the output $\boldsymbol{y}^t = \left\{y_1^t, \cdots, y_{L_y}^t \middle| y_i^t \in \mathbb{R}^{d_y}\right\}$.

The encoder-decoder architecture is shown in Figure 1. The encoder takes a variable-length feature vector $\boldsymbol{x}^t$ as input and converts it into a fixed-dimensional encoding state $\boldsymbol{h}^t = \left\{h_1^t, \cdots, h_{L_h}^t \middle| h_i^t \in \mathbb{R}^{d_h}\right\}$. The decoder maps the fixed-dimensional encoding state $\boldsymbol{h}^t$ to a variable-length feature vector $\boldsymbol{y}^t$.

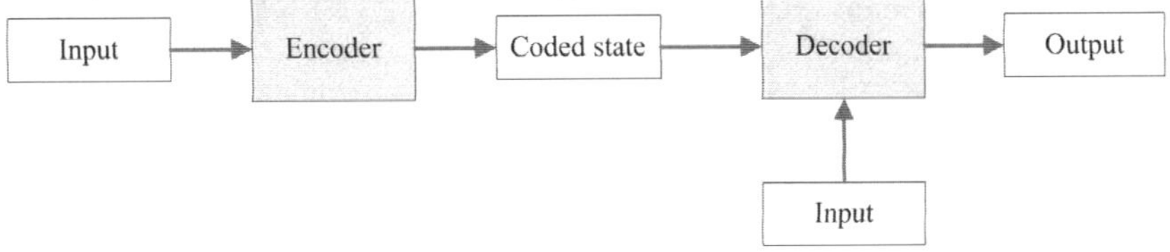

Figure 1. Encoder-decoder architecture

4. The proposed method

4.1. Model input

Since the supply of hydraulic systems is affected by the load of hydraulic supports, and the actions of hydraulic supports follow the movement of the shearer, the input parameters of the model include the position of the shearer, the column pressure of the hydraulic support in front and behind the shearer, the forward slip travel, the action execution type, the action encoding, whether the action is in progress, and the controller status, etc. Since all the parameters included in the input have sequential characteristics, the position information of each time step in the input sequence needs to be assigned to the network model to make the model understand the order relationship of the input sequence. The Transformer Neural Network first performs word embedding on each data in the input sequence. Word embedding elevates the original one-dimensional data to a two-dimensional matrix and maps each value in the input sequence to a fixed-dimensional feature row vector. Then, the Transformer Neural Network encodes the input sequence with sine and cosine functions, generates a fixed absolute position representation, i.e., positional encoding, and adds it to the previously completed word embedding sequence in a positional manner [10]. The formula for positional encoding is as follows:

$$
\begin{aligned}
PE(pos, 2i) &= \sin\left(pos / 10000^{\frac{2i}{d_{model}}} \right) \\
PE(pos, 2i+1) &= \cos\left(pos / 10000^{\frac{2i}{d_{model}}} \right)
\end{aligned}
\tag{1}
$$

where *pos* represents the position index of a certain time step data in the input sequence; d_{model} is the dimension of the input sequence word embedding; i is a certain vector dimension.

4.2. Model structure

The self-attention mechanism can make the model pay more attention to the intrinsic relationship between the input sequence data, which is the core technology of the Transformer neural network. Its calculation process is defined in Equation (2) as follows:

$$
\text{Attention}(\boldsymbol{Q}, \boldsymbol{K}, \boldsymbol{V}) = \text{Softmax}\left(\frac{\boldsymbol{Q}\boldsymbol{K}^T}{\sqrt{d_{model}}} \right)\boldsymbol{V} \tag{2}
$$

where Q represents the query; K represents the key; V represents the value, and d_{model} is used as the scaling factor for the word embedding dimension, making the gradient more stable during the training process.

The schematic diagram of the Liquid Demand Prediction Model based on the Transformer is shown in Figure 2. For the model's encoder, the multi-head attention mechanism is mainly used to search for important information in different subspaces and integrate information to enhance the feature representation ability and improve the model's prediction accuracy. For the decoder based on the self-attention mechanism, as the entire sequence is exposed, the input sequence of the decoder needs to be masked to hide the relevant information of the future time step. Therefore, the decoder is mainly based on a masked multi-head self-attention mechanism, and the calculation process is roughly the same as that of the encoder.

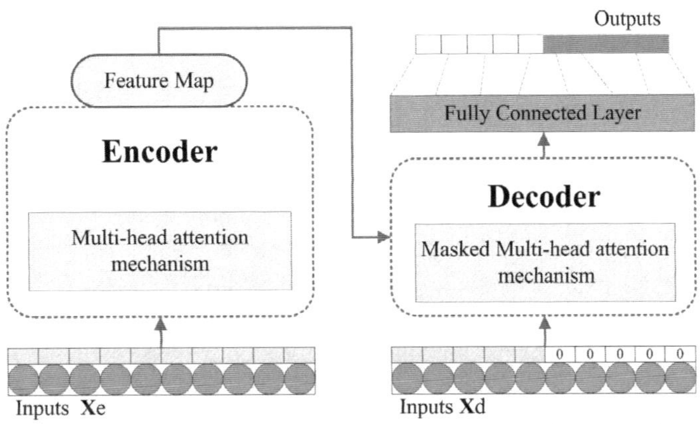

Figure 2. Transformer-based Liquid Demand Prediction Model

4.3. Liquid demand prediction

As shown in Equation (3), Mean Square Error (MSE) was selected as a loss function on the target sequence of liquid demand to guide the optimization of model parameters.

$$MSE = \frac{1}{m}\sum_{i=1}^{m}\left(y_i - \widetilde{y_i}\right)^2 \tag{3}$$

The feature vectors are constructed based on the monitoring parameters of the coal mining machine's position, hydraulic support, and action parameters. They are divided into training, validation, and test sets. A Transformer Neural Network Prediction Model is constructed, and parameters and hyperparameters are set. The model is trained using the training set, and the model training effect is validated using the validation set. A certain duration of historical samples from the test set is taken as input to the Transformer Hydraulic Liquid Demand Prediction Model. The model predicts the future hydraulic liquid demand for a certain duration, providing information for the intelligent control of the hydraulic system.

4.4. Intelligent supply control based on predicted liquid demand

Currently, coal mine supply systems mainly use multiple parallel emulsion pumps, each with an electromagnetic unloading valve to control the outlet pressure of a single pump. At the same time, the speed of the driving motor can be adjusted through a frequency converter to change the supply of the emulsion pump. By combining electromagnetic unloading and frequency conversion speed regulation, the supply flow rate can be continuously adjusted to provide the required power for the hydraulic support. In addition, for a multi-pump supply system, the start-up sequence of the emulsion pumps is generally set as the main pump, secondary pump, auxiliary pump, and backup pump to achieve multi-pump linkage and facilitate dynamic adjustment of the system supply. Only when a certain emulsion pump is manually stopped or stopped due to a failure is the sequence of emulsion pump grouping readjusted.

Based on the predicted hydraulic demand model, the prediction of booster unloading and hydraulic support actions at the pumping station can be achieved. Guided by the predicted hydraulic demand results, relying on the "main, secondary, auxiliary, backup" multi-pump linkage mechanism and using frequency conversion and electromagnetic unloading as adjustment methods, intelligent supply control can be achieved according to demand, as shown in Figure 3.

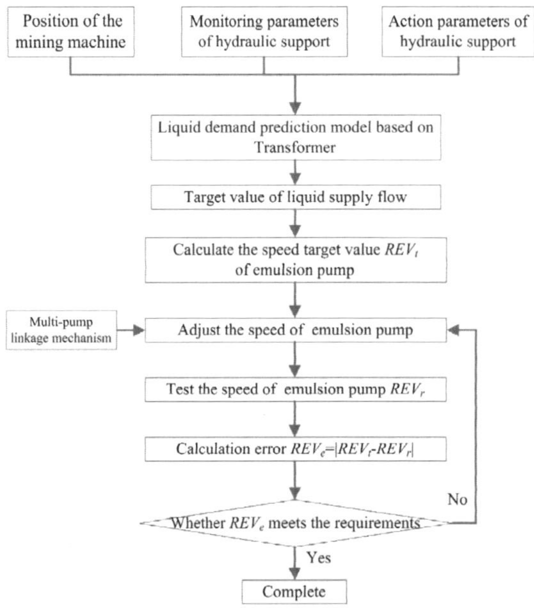

Figure 3. Intelligent control flow of liquid supply on demand

5. Experimental verification

5.1. Long-distance liquid supply system and setting of the model

To verify the stability and response of the method proposed in this paper, monitoring data from Huangling Coal Mine were used for application verification. The coal mining face of Huangling Coal Mine has 156 hydraulic supports. To better monitor the flow rate and pressure required by the hydraulic supports, flow sensors are installed in the measurement system to measure the flow rate provided to the hydraulic supports. Pressure sensors are installed in the measurement system every 25 hydraulic supports to measure the pressure. Based on the position of the coal mining machine, a total of 99 parameters, including the pillar pressure, front travel distance, action execution type, action code, whether the action is in progress, and controller status of the 9 hydraulic supports before and after the current position, were considered. Therefore, the input sample contains 100 parameters.

Relevant monitoring data were collected through the LongWallMind system, and 80% of the data were used as the training set, 10% of the data were used as the validation set to check the training effect of the model, and the remaining 10% of the data were used as the test set to test the prediction accuracy of the trained model.

For the Transformer-based Liquid Demand Prediction Model, a 6-layer perceptron (128-128-256-256-512-512) is used to construct the encoder, and a 3-layer perceptron (128-256-256) is used to construct the decoder. The input dimension of the encoder is 100. Four attention matrices are used in the multi-head attention layer to project the features into four different feature spaces. The model training uses a dynamic learning rate and sets the initial learning rate for model training to 0.001. After each training round, the learning rate is multiplied by 0.8 before starting a new round of training to accelerate the training process while ensuring that the model can converge.

5.2. Liquid demand prediction

To fully explore the potential of the Transformer-based Liquid Demand Prediction Model for predicting the required flow rate and pressure of hydraulic supports, the model is used to predict the liquid demand for coal-cutting machine movements at different distances of (1, 2, 3, 4), whose periods are corresponding to about (18, 36, 54, 72) seconds, with 10 predictions made for each duration. The prediction accuracy of normalized flow and pressure measured in MSE is shown in Figures 4 and 5.

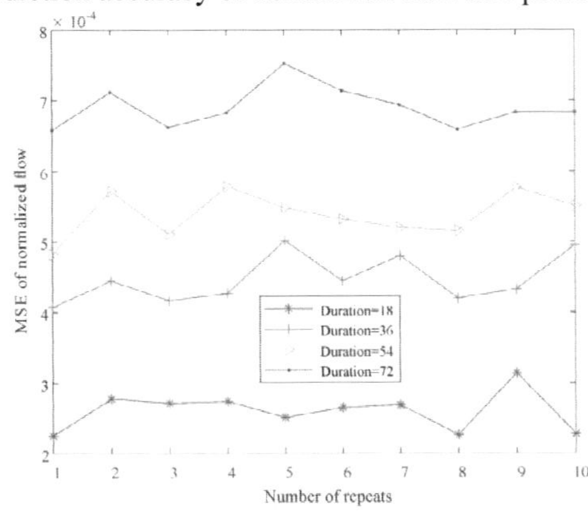

Figure 4. MSE of normalized flow

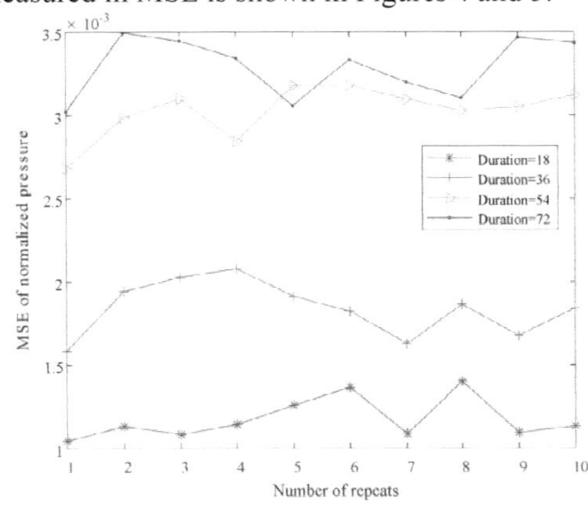

Figure 5. MSE of normalized pressure

It can be seen from Figures 4 and 5 that the accuracy and performance of the Liquid Demand Prediction Model for predicting flow and pressure are quite high and stable. In addition, although the model's prediction error increases with the prediction time, the trend is very slow. Therefore, the Liquid Demand Prediction Model can fully capture the long-term dependency features in the liquid demand for hydraulic supports and make accurate predictions. While doubling the prediction time, it can effectively control the growth of errors, adjust the liquid supply strategy in advance, and respond promptly to the action of hydraulic supports.

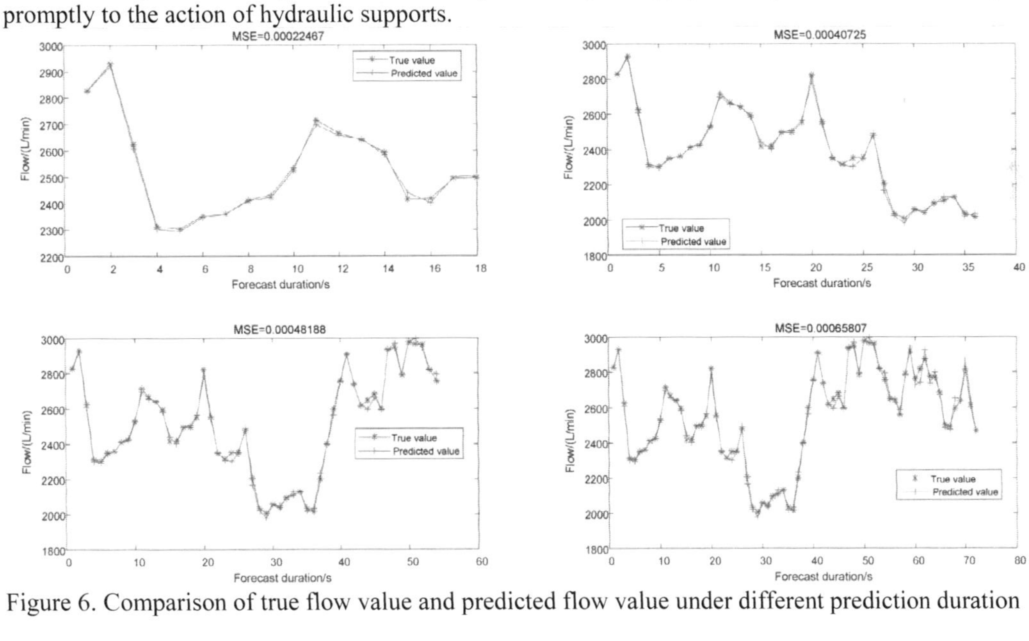

Figure 6. Comparison of true flow value and predicted flow value under different prediction duration

The normalized flow and pressure are restored to demonstrate the model's prediction accuracy intuitively. And the true value and predicted value comparison of flow and pressure under different prediction duration are shown in Figures 6 and 7. Although the accuracy of predicting liquid demand may be lacking at the peak of fluctuations, the prediction model can predict the trend of liquid demand fluctuations well regardless of the length of prediction time, and there is no obvious time delay effect.

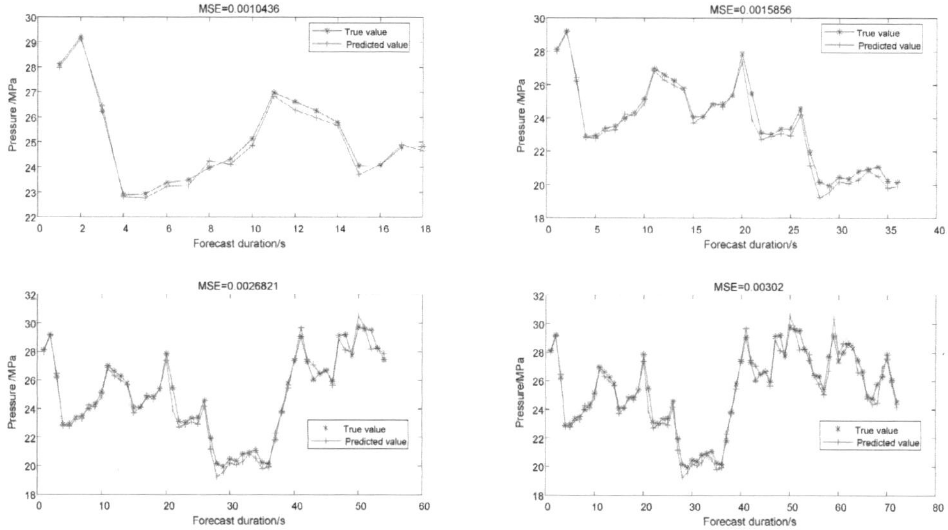

Figure 7. Comparison of true pressure value and predicted pressure value under different prediction duration

6. Conclusion

A long-distance intelligent liquid supply method for coal mining faces based on liquid demand prediction is proposed in this paper. The method uses a Transformer Model to establish a Liquid Demand Prediction Model for hydraulic supports. It solves the problem of long sequence time prediction problem by combining attention mechanisms to predict liquid demand accurately. Furthermore, based on the predicted liquid demand, it combines multiple pump linkage mechanisms to achieve intelligent liquid supply and meet the liquid demand prediction requirements of hydraulic supports under different working conditions, which can ensure the speed and accuracy of hydraulic support movement, as well as the effectiveness of the work efficiency and other indicators.

References

[1] Wang G and Liu F 2020 *China Coal Mine Intelligent Development Report* (Beijing: Science Press)

[2] Li R, Liu B, Wang D, *et al.* 2022 Progress and application of intelligent fluid supply technology in working face *Coal Sci. Technol.* **50** 247-253

[3] Zhang N, Lang R, Wang D, *et al.* 2023 Coalface long-distance fluid for key technology present situation and prospect *J. of Coal Proj.* **55** 71-75

[4] Fu X, Wang R, Zhao Y, *et al.* 2020 Self-adaptive control method of fluid feeding with steady pressure for hydraulic support based on overlapping synergetic logic *J. of China Coal Soc.* **45** 1891-1900

[5] Si M, Wu B, Wang Z 2022 Research on large flow intelligent liquid supply system in fully mechanized working face *J. of Mine Automation* **48**: 66−72

[6] Peng Z Y, Guo P C 2022 A data organization method for LSTM and transformer when predicting chinese banking stock prices *Discrete Dynamics in Nature and Society* **2022**: 1−8

[7] Cui H, Zhou K, Hu F, *et al.* 2023 Power equipment state fusion prediction model based on improved LSTM *Elect. Meas. Instr.* **60**: 10-15

[8] Qi X Y, Hou K, Liu T, *et al.* 2021 From known to unknown: knowledge-guided transformer for time-series sales forecasting in alibaba. *arXiv*

[9] Giuliari F, Hasan I, Cristani M, *et al.* 2020 Transformer networks for trajectory forecasting *25th Int. Conf. on Pattern Recog.* 10335-10342

[10] Vaswani A, Shazeer N, Parmar N, *et al.* 2017 Attention Is All You Need. *arXiv*

Uninsulated Track Circuit Parameter Adjustment and Ballast Resistance Analysis

Tengfei Ma[1]*, Yunshui Zheng[1], Jing Xu[2], Jianfa Wang[2]

[1] School of Automation and Electrical Engineering, Lanzhou Jiaotong University, Lanzhou, 730070, Gansu, China.

[2] Guangzhou Electrical Services Section, China Railway Guangzhou Group Co., Ltd. Guangzhou, 510088, Guangdong, China.

*mail: 12211509@stu.lzjtu.edu.cn

Abstract: Uninsulated track circuits are increasingly used in railroad signaling equipment, but the external environment dramatically influences operating conditions. Using the relationship between uninsulated track circuits and ballast resistance, an equivalent four-terminal network model of track circuit transmission characteristics is constructed based on uniform transmission line theory to analyze the influence law of ballast resistance on track surface current. It also makes corresponding adjustments to the track circuit parameters to ensure that the track circuit can work reliably in all four different operating conditions to ensure the safety of railroad transportation.

1. Introduction

The track circuit is the core component of the train control system, which transmits traffic information to the train through the rails and is the essential equipment to reflect the running status of the train and check the idle occupation of the zone, which plays a crucial role in ensuring the safety of train operation and improving transportation efficiency.[1] Ballast resistance is a leakage resistance between two rails; it is susceptible to the climate conditions of the railroad site (temperature, humidity), the quality and state of ballast, the quantity and thickness of ballast, the type of rail sleeper and insulation performance, wooden rail sleeper used in the composition of the preservative and corrosion treatment methods. It can be seen that the change in ballast resistance is significant for the safety of traffic, so we need to study reasonable measurement methods to realize the accurate measurement of ballast resistance. The original measurement methods, such as the open circuit short circuit method,[2] three impedance method, phase meter method [3], etc., are mainly for insulated track circuits, while most of the domestic high-speed railroads use uninsulated track circuits, which have compensation capacitance distributed between rails at intervals; therefore, the above methods are not applicable. An accurate and reasonable ballast resistance measurement method is needed for the widely used uninsulated track circuits.[4] In this paper, based on the transmission principle of rail circuits, we study the influence law of ballast resistance on rail surface current, combine the measured rail surface current data in the field, build a data set, and establish a ballast resistance regression prediction model to realize the accurate measurement of ballast resistance.

To summarize, many factors affect ballast resistance, and its resistance value can easily change, which in turn affects the regular operation of the track circuit. Therefore, to ensure the reliable operation of the track circuit and the quality of the circuit signal transmission and to improve the efficiency of maintenance personnel, it is of great importance to study how to adjust the track circuit scientifically and efficiently when the ballast resistance changes within the specified range.[5]

2. Structure and transmission principle of orbital circuits

The uninsulated track circuit is composed of a transmitter (FS), cable simulation network (ML), matching transformer (BP), tuning unit (BA), open core coil (SVA), compensation capacitor (C), tuning unit (BA), matching transformer (BP), cable simulation network (ML), depleter (S), receiver (JS), and track relay (GJ) from the transmitting end to the receiving end respectively. The uninsulated track circuit divides the original track into the primary track circuit and the small track circuit in the tuning area. The small track circuit is regarded as the "continuation section" of the introductory track circuit ahead of the train operation. According to the structure of the uninsulated track circuit, it is divided into five areas: receiver tuning area, transmitter tuning area, receiver channel, transmitter channel, and main track line. Its general structure is shown in Fig.1. The transmitter will shift the frequency signal through the transmitter channel, the transmitter side of the tuning area, along the main track line to the receiver side of the tuning area, and then through the receiver channel to the indoor attenuator and finally reach the receiver, the receiver judgment to drive the track relay suction up and down, to determine the idle and occupied situation of the track section.

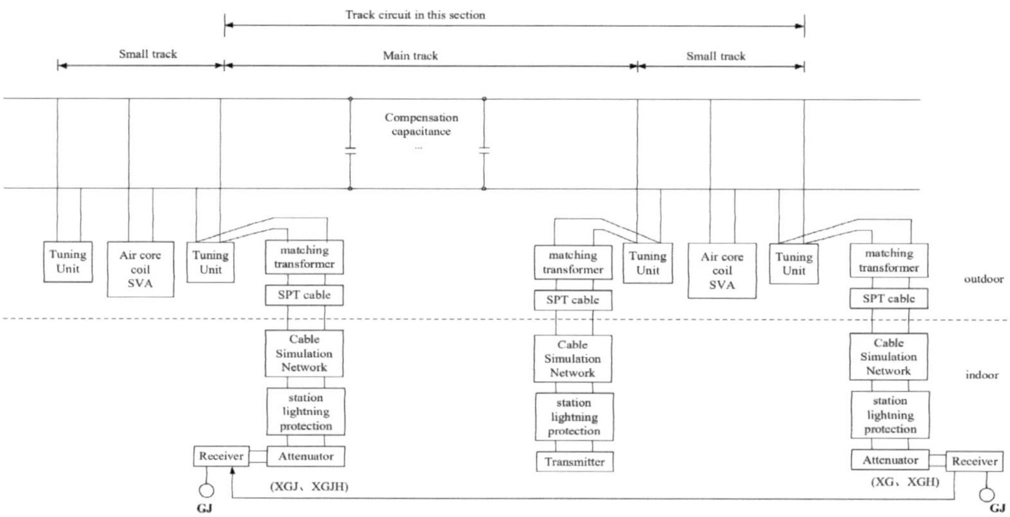

Figure 1. Schematic diagram of orbital circuit system composition

Because of the existence of voltage differences between the two rails, the track circuit's electric energy in the transmission process results in part of the signal current from one rail flow through the rail sleeper, ballast to another rail leakage, and affects the signal stability transmission. The leakage current is uniformly distributed along the track line at all points, so the track circuit in the communication of electrical energy belongs to the uniform transmission line. The voltage at each point along the line is not distributed according to a linear distribution law but according to the law of close hyperbolic function down. Because in each unit length, there is leakage current, so as the transmission distance increases, the track current gradually decreases, and the voltage also gradually decreases. Ballast resistance, the smaller the leakage current between the two rails, the larger the track circuit power consumption is also more significant, resulting in the transmission to the receiver signal current being too low, resulting in a "red band" fault; [6] Conversely, the greater the ballast resistance, the leakage current between the two rails is smaller, resulting in the transmission to the receiver's residual voltage is too high, resulting in "poor splitting" fault. And the greater the range of ballast resistance changes, the more unstable the track circuit work, quickly affecting the safety of traffic.

3. Current equivalent model

From the previous paper, it is clear that the magnitude of ballast resistance affects the change of track surface current in the track circuit adjustment state. To analyze the influence of ballast resistance on the track circuit, the track circuit is modeled according to uniform transmission line theory and a four-terminal network, and the influence of ballast resistance on the track surface current is simulated and analyzed.

In the adjustment state, the rail surface is randomly taken to a point x, let its voltage $\vec{U}(x)$, current $\vec{I}(x)$ respectively, $\vec{Z}_j$ is the rail surface receiving endpoint to the receiving end of the apparent incoming impedance, $\vec{Z}_{(x)}$ is any point x to the receiving end of the evident incoming impedance, $\vec{N}_j(x)$ is the receiving end to any point x of the track circuit transmission equivalent four-terminal network, $\vec{N}_f(x)$ is the transmitter to any point x of the track circuit transmission identical four-terminal network. Then the comparable model of the current at any point x on the rail surface is shown in Fig. 2.

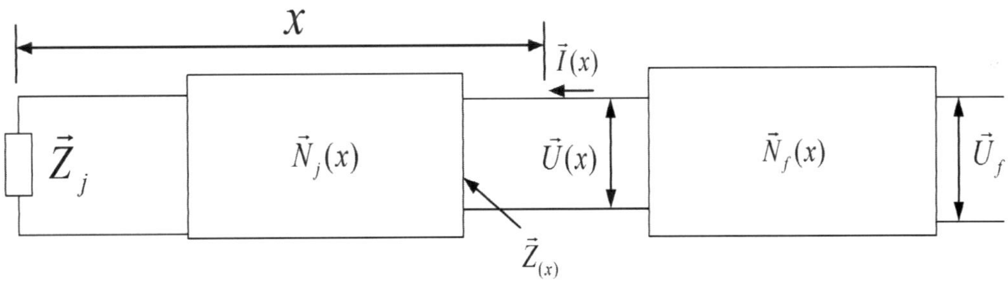

Figure 2. Rail circuit rail surface current equivalent model

From the theory of uniform transmission lines [8], it follows that the current $\vec{I}(x)$ at any point x on the rail surface is expressed as:

$$\left|\vec{I}(x)\right| = \frac{A_f}{\left|\vec{N}_{f11}(x)\vec{Z}_{(x)} + \vec{N}_{f12}(x)\right|} \tag{1}$$

Where A_f is the amplitude of the signal sent by the transmitter and is a constant value.

The apparent incoming impedance $\vec{Z}_{(x)}$ at any point x on the rail surface toward the receiving end can be expressed as

$$\vec{Z}_{(x)} = \frac{\vec{N}_{j11}(x)\vec{Z}_j + \vec{N}_{j21}(x)}{\vec{N}_{j12}(x)\vec{Z}_j + \vec{N}_{j22}(x)} \tag{2}$$

According to the working principle of rail circuit signal transmission, it is known that

$$\vec{U}(x) = \vec{I}(x)\vec{Z}_{(x)} \tag{3}$$

$$\vec{U}_f = \vec{N}_{f11}(x)\vec{U}(x) + \vec{N}_{f12}(x)\vec{I}(x) \tag{4}$$

Equation (2), Equation (3), and Equation (4) can be solved in conjunction to obtain Equation (1), and finally, the rail surface current model can be determined as Equation (1).

4. Effect of ballast resistance on track circuits

According to the parameter setting in the track circuit adjustment table, the cable length is 10 km, the signal carrier frequency is 1700 Hz, the track circuit length is 1300 m, and the transmit level is three as the base parameter because the ballast resistance takes the value of the lower limit of 0.5Ω·km and is only possible in terrible weather; when the ballast resistance value is of more than 10Ω·km, the value of the rail surface current makes no change, very close to the resistance value infinity, so, set the ballast resistance value of 0.5, 1, 2, 3, 4, and 10 Ω·km, every interval of 100 meters to calculate the corresponding rail surface current, the results is shown in Fig. 3.

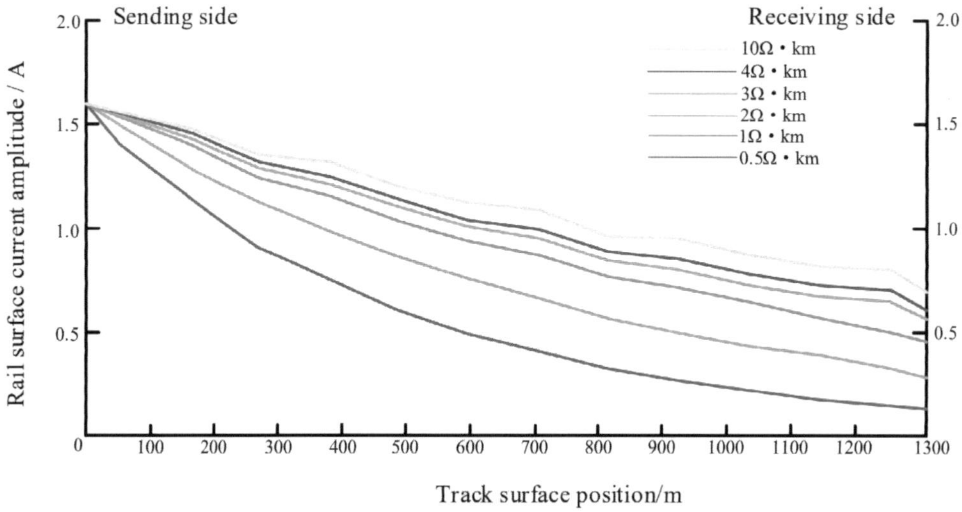

Figure 3. Rail surface current values at different ballast resistances

From the figure we can see that from the transmitting end to the receiving end, with the increase of the unit track circuit length, the track surface current overall is similar to the exponential decay trend, and the degree of decay of the track surface current with the ballast resistance increases and decreases; different ballast resistance has other effects on the track surface current, with the ballast resistance increases, the track surface current decay is more minor and gradually tends to be smooth.

5. Orbital circuit adjustment

From the above analysis, it can be seen that the ballast resistance has an impact on the track surface current at any point of the track circuit, mainly in the degree of attenuation, so to ensure the reliable operation of the track circuit, the track circuit needs to be adjusted for changes in the ballast resistance.

The adjustment of the uninsulated track circuit includes the adjustment of transmitter carrier frequency, analog cable network cable compensation length adjustment, transmit level adjustment, primary track signal adjustment, and minor track signal adjustment in positive and negative directions.[9] The transmitter carrier frequency adjustment and analog cable network cable compensation length adjustment before the line is opened, the project construction should be adjusted according to the actual situation of the bar combined with the design standards and combined with the wiring diagram for verification and confirmation. In the later maintenance process, the carrier frequency and analog cable compensation are generally no longer adjusted.

The transmit power out level adjustment, primary track signal adjustment, and positive and negative minor track signal adjustment must be based on the track circuit adjustment table. According to the actual ballast resistance of the track circuit to make the corresponding adjustment, the track circuit adjustment table is based on the total cable length, carrier frequency, section length range, and the lowest ballast resistance to be calculated separately. If the ballast resistance value is less than the minimum ballast resistance specified in the adjustment table, the leakage is too significant. To prevent the track circuit "red light band", the receiving level is needed to be appropriately increased to meet the requirements of more significance than the minimum ballast resistance value; if the ballast resistance tends to infinity, to prevent the residual voltage above the tap and "poor", the reception level can be appropriately lowered to meet the requirements of less than the maximum value. The adjustment of the parameters of the track circuit must be strictly under the adjustment table to implement, not arbitrarily adjusted. Blindly raising or lowering the transmit power out level, the main track receives level will lead to the transmission of the track circuit, matching characteristics becoming bad, and directly affecting the normal working state of the track circuit.[10]

6. Conclusion

This paper constructs an equivalent four-terminal network model of the transmission characteristics of the track circuit under the adjustment state of the uninsulated track circuit. It obtains the influence law of the ballast resistance on the current amplitude of the track surface through simulation analysis, which provides a theoretical basis for the subsequent adjustment of the track circuit. The relationship between ballast resistance and track circuit is analyzed and mastered, the actual situation of ballast resistance is analyzed through testing, and the arrangement of the uninsulated track circuit is adjusted accordingly and verified. To accurately grasp the working state of the track circuit can improve the measurement efficiency and also has essential significance for field maintenance.

References

[1] Zhao L, Shi W."Modeling and Simulation of Information Transmission Process from JTC to TCR Based on the Finite Element Method and Transmission Line Theory." Journal of the China Railway Society 35.08(2013):63-69. DOI:10.3969/j.issn.1001-8360.2013.08.010
[2] Yuan X. "Study on Faulty Track Circuit Branch." Railway Signalling & Communication. 04(2007):11-14. DOI:10.13879/j.issn1000-7458.2007.04.005.
[3] Yang K, Zhang X, Li D. "Way for Measuring Track Circuit Parameter on Line." Railway Signalling & Communication. 12(1997). DOI:10. 1387 9/j.issn1000-7458.1997.12.002
[4] Zhao L, Liu C, Ran Y. "On-Line Monitoring of Ballast Resistance of Jointless Track Circuits." Journal of the China Railway Society 39.08(2017):101-106. DOI:10.3969/j.issn.1001-8360.2017.08.014
[5] Cui Y, Zhao L, Xu X. "Fault Diagnosis Method for JTC Rail Surface Equipment Based on Stock Analysis and Grey Correlation Analysis." Journal of the China Railway Society 43.05(2021):112-120. DOI: 10. 3969 /j.issn.1001-8360. 2021. 05. 013
[6] Geng Y. "Discussion on Several Problems for Ballast Resistance of Track Circuit." Railway Signalling & Communication Engineering 16.02(2019):72-75. DOI: 10.3969/j.issn.1673-4440.2019.02.017
[7] Sun S, Zhao H, Quan H, Chen D, Lin T, Ning B. "Equipment Fault Diagnosis for Electrical Separation Joint of Jointless Track Circuits Using Qualitative Trend Analysis." China Railway Science 35.01(2014):105-113. DOI:10.3969/j.issn.1001-4632.2014.01.17
[8] Hill, R. J., S. Brillante, and P. J. Leonard. "Electromagnetic field modeling for transmission line distributed parameters of railway track." IEE Proceedings-Electric Power Applications 146.1 (1999): 53-59, DOI:10.1049/ip-epa:19990193
[9] Huang M. "Study on Principle and Method of ZPW-2000A Track Circuit Adjustment." Application of IC 38.10(2021):279-281. DOI:10.19339/j. issn.1674-2583.2021.10.133.
[10] Jin Y. "Discussion on Adjustment Method of 25 Hz Phase Sensitive Track Circuit." Telecom Power Technology 36.06(2019):35-36. DOI:10.19399/j.cnki.tpt.2019.06.013.

Three-level ANPC rectifier loss calculation and balance control method

Zi Ying[1, *]

[1] School of Electrical and Control Engineering, North China University of Technology, Shijingshan District, Beijing, China

[*]815908217@qq.com

Abstract. Traditional three-level neutral-point clamped converter has the problem of the uneven loss distribution, which restricts the increase of converter capacity. Three-level active neutral-point clamped (3L-ANPC) converter can reduce the imbalance of power device loss distribution. At present, three traditional SPWM modulation strategies are used in single-phase three-level ANPC modulation. Considering the loss distribution, it is generally required to input the current and voltage sampling values in real-time and calculate the temperature of the switching device online for control or collect the temperature of the switching device in real-time for feedback control, which is more complicated to realize. This paper proposes a modulation scheme based on two carrier cycles, combined with traditional SPWM1 and 2, to analyze the switching conditions and loss characteristics of each switching device in this scheme. The simulation of a single-phase three-level ANPC rectifier was built, and the experiment was carried out on the Yuankuan HIL semi-physical platform. The results show that this scheme is simple, easy to implement, and can effectively reduce the loss distribution imbalance.

1. Introduction

Compared with the three-level NPC topology, the three-level ANPC topology can achieve loss balance by selecting an appropriate zero-level output state and changing the on-off and off-off conditions of switching devices. For three-level ANPC converters, much literature at home and abroad put forward different modulation methods. However, in the practical application, SPWM is still mainly adopted. Its loss balance control method generally requires the online real-time calculation of switching device temperature, direct control, or feedback control, which greatly increases the amount of computation; the realization process is more complex [1].

To solve the above problems, this paper proposes a modulation scheme that takes two carrier cycles as a unit and synthesizes traditional SPWM1 and 2. Without increasing the complexity of control, the total loss is between strategy 1 and strategy 2, which is significantly less than the total loss of strategy 3. At the same time, the switching times of each power device are more average, and the unbalance degree is significantly reduced.

2. Topological structure and analysis

2.1. Topological structure and analysis

The main circuit topology of the single-phase three-level ANPC rectifier is shown in Figure 1.

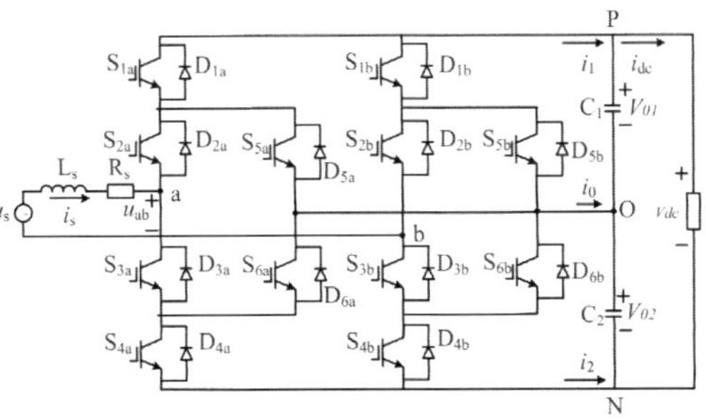

Figure 1. Topology of ANPC three-level rectifier.

Each phase bridge arm of a three-level ANPC rectifier consists of six IGBTs. $S_{1x} \sim S_{6x}$ represents 6 IGBT switches of the same model, $D_{1x} \sim D_{6x}$ represents six clamp diodes, and C_1, C_2 represents DC side voltage divider capacitors with the same parameters. Among them, S_{1x}, S_{4x}, D_{1x} and D_{4x} are referred to as outer layer devices, S_{2x}, S_{3x}, D_{2x}, D_{3x} are referred to as inner layer devices, and S_{5x}, S_{6x}, D_{5x}, D_{6x} are referred to as midpoint clamping devices [2].

Due to the action of the active clamp circuit, each phase bridge arm can output P, O, and N three level states. In the O state, current can flow in or out through S_{2x} and S_{5x} of the upper bridge arm or S_{3x} and S_{6x} of the lower bridge arm. Therefore, the O state can be divided into OU1, OU2, OL1, and OL2 paths [3]. Thus, the three-level ANPC rectifier has six switching states per phase.

2.2. Traditional modulation method

There are three typical SPWM modulation strategies.

Using modulation method 1, when the reference voltage $U_{ref} > 0$, the converter switches between P and OU1. When the reference voltage $U_{ref} < 0$, the converter switches between N and OL1. The switching frequency of S_{5x} and S_{6x} is consistent with the system switching frequency. The switching frequency of S_{1x} and S_{4x} is half of the system switching frequency. The switching frequency of S_{2x} and S_{3x} is consistent with the frequency of modulation voltage. Switching loss is concentrated in the outer layer and the midpoint clamping device [4].

Using modulation method 2, when the reference voltage $U_{ref} > 0$, the converter switches between P and OL2. When the reference voltage $U_{ref} < 0$, the converter switches between N and OU2. The switching frequency of S_{2x} and S_{3x} is consistent with the system switching frequency. The switching frequency of S_{1x}, S_{4x}, S_{5x}, and S_{6x} is consistent with the modulation voltage frequency. Switching losses are concentrated in the inner layer devices.

Under modulation method 3, the switching frequency of S_{2x}, S_{3x}, S_{5x}, and S_{6x} is consistent with the system switching frequency. The switching frequency of S_{1x} and S_{4x} is half of the system switching frequency. The distribution of switching loss in each component is relatively average, but the total switching loss is significantly increased compared with methods 1 and 2 [5].

2.3. New modulation strategy and loss analysis

Under the condition of modulation voltage frequency of 50 Hz and system switching frequency of 10 kHz, the analysis shows that within a modulated wave cycle of 0.02 s, the six power devices in each phase of method 1 are $2\times200+2\times100+2\times2=604$ times, and the six power devices in each phase of method 2 are $2\times200+4\times2=408$ times. For method 3, there are $4\times200+2\times100=1000$ times.

Two modulated wave periods are taken as the unit, and methods 1 and 2 are adopted alternately to modulate.

In order to calculate the switching and conduction losses of power semiconductor devices in simulation, it is necessary to establish a loss model for quantitative analysis.

The data manual provided by Infineon provides the voltammetry characteristic curve between the voltage and the current flowing on both sides of the IGBT and diode under normal working conditions. The relationship curves between switching loss E_{on}, E_{off} and current I_c of IGBT devices, and the relationship curves between diode reverse recovery loss E_{recD} and current I_F are also given.

2.4. Loss model

IGBT loss consists of conduction loss and switching loss.

$$P_{loss} = P_{con} + P_{sw} \tag{1}$$

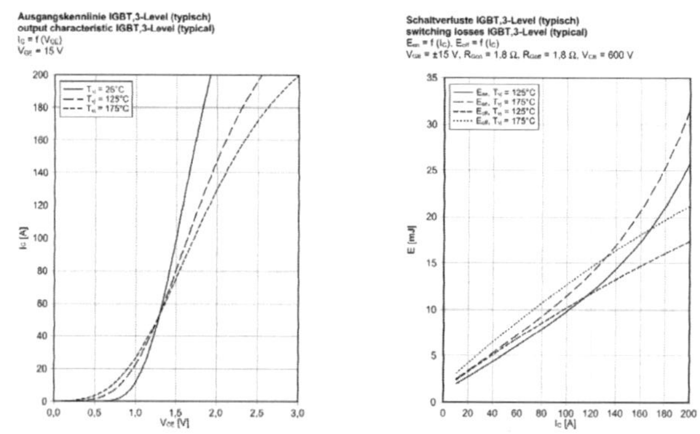

(a) (b)

Figure 2. IGBT loss characteristic curve

According to the IGBT volt-current characteristic curve shown in Figure 2(a), this nonlinear relationship is usually modeled in a rough approximation as a voltage source and resistance in series with an ideal switch. In general, the current I_c is directly multiplied by the corresponding voltage U_{CE} to obtain the conduction loss $P_{V,COND}$ depending on the current I_c [6][7]. The calculation expression of conduction power at 125 °C can be obtained by fitting Figure 2.

$$P_{V,COND,TA=125} = c_{TA} I_c + d_{TA} I_c^2 = 0.8370 I_c + 0.0081 I_c^2 \tag{2}$$

By applying the actual current through IGBT to the second-order polynomial fitting function, the conduction power of IGBT can be calculated directly.

According to the data manual provided by Infineon Company, the relationship between IGBT primary turn-on or turn-off loss and current is shown in Figure 2(b). The calculation expression of primary switching loss under 125 °C operating conditions is obtained:

$$E_{TOTAL} = \left(a + bI_c + cI_c^2\right) I_c = \left(0.3995 - 0.0074 I_c + 0.0001 I_c^2\right) I_c . \tag{3}$$

Similarly, diode loss is composed of conduction loss and reverse recovery loss.

$$P_D = P_{conD} + P_{recD} \tag{4}$$

According to the voltammetry characteristic curve of the diode, the formula for calculating the conduction power at 125 °C can be obtained:

$$P_{conD} = -0.0001 I_f^3 + 0.0185 I_f^2 + 0.3196 I_f . \tag{5}$$

According to the reverse recovery characteristic curve of the diode, the calculation expression of reverse recovery loss at 125 °C can be obtained:

$$E_{recD} = -0.0001 I_f^{\ 2} + 0.0603 I_f + 0.9267\,. \tag{6}$$

The total conduction loss power and the total switching loss of the diode in one cycle are obtained [8].

3. Simulation and experimental results

3.1. Simulation results and analysis

In order to verify the effectiveness of the proposed method, a simulation model of a single-phase three-level ANPC rectifier was built in Simulink. And experiments were conducted on Yuankuan HIL semi-physical platform. The specific simulation parameters are shown in Table 1.

Table 1. Parameters of simulation and experiment.

Parameters	*Value*	Parameters	*Value*
Rated AC voltage	120 V	Rated frequency	50 Hz
Rated DC voltage	150 V	Rated Capacitance	16 mF
Rated inductor	2 mH	Switching frequency	10 kHz
Rated resistance	0.15 Ω		

In order to quantify the loss unbalance degree of n switching device modules in the system, the calculation expression of unbalance degree function is defined as:

$$f_{un} = \frac{\max\left(P_{TjDj}\right) - \sum_{j=1}^{n} \frac{P_{TjDj}}{n}}{\sum_{j=1}^{n} \frac{P_{TjDj}}{n}}\,. \tag{7}$$

Under the same switching frequency and PF=1, the four types of losses of six A-phase power devices are simulated respectively in methods 1, 2, and 3, and the proposed rotation modulation method. The total A-phase loss of the four modulation methods is calculated. The power device module loss imbalance of the four modulation methods is calculated and compared according to Formula (7).

Figure 4 shows the total A-phase loss data and the module loss unbalance f_{un} of switching devices calculated by four modulation methods based on Formula (7).

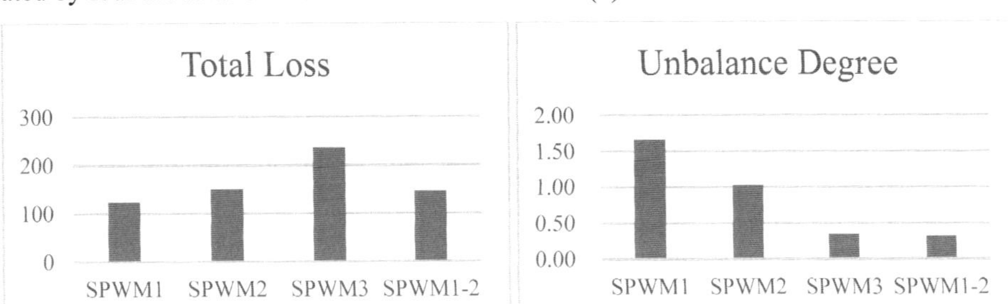

Figure 3. Total A-phase loss and unbalance degree.

According to Figure 3, under unit power factor rectification, the rectifier adopts the control method selected in this paper. The unbalance degree is 0.32, which is basically consistent with the unbalance degree of traditional method 3 and much lower than that of traditional method 1 and traditional method 2. Compared with the total A-phase loss of switching devices under the four types of modulation methods in Figure 3, the total loss of traditional method 3 is significantly greater than that of traditional method 1 and method 2 due to its greatly increased number of breaks. The method in this paper is essentially a combination of traditional method 1 and method 2, so the total loss is also between the total loss of the two [9][10]. To sum up, although the unbalance degree of traditional method 3 is basically the same as that of the method in this paper, and both can achieve a good loss balance effect, the total loss of traditional method 3 is 1.63 times that of the total loss of the method in this paper. The method in this paper has a better comprehensive effect.

3.2. Experimental results and analysis

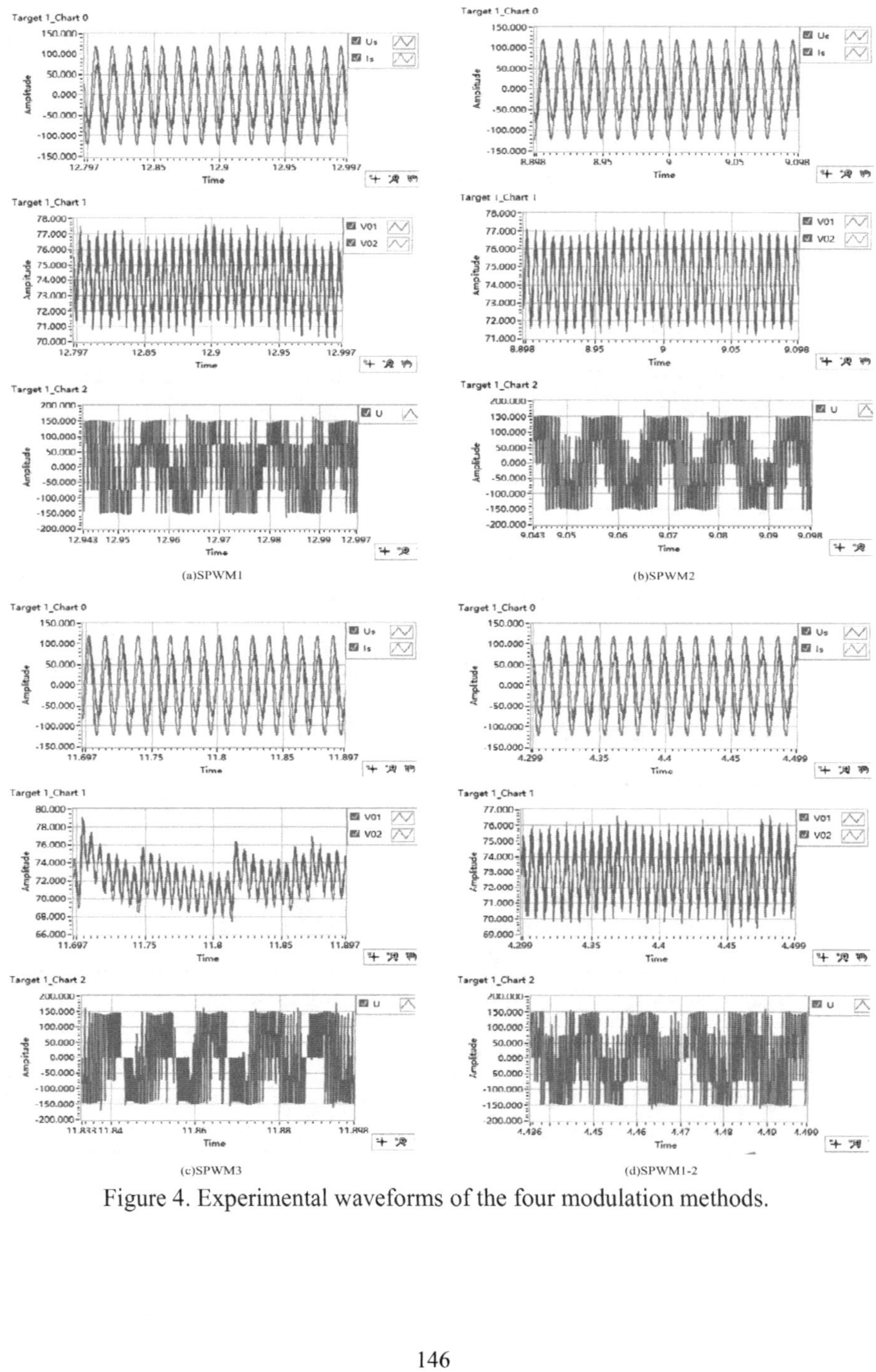

Figure 4. Experimental waveforms of the four modulation methods.

The experimental parameters are chosen to be consistent with the simulation, and the steady-state experimental waveforms of the four modulation methods are shown in Figure 4. Us and Is are the waveforms of network side voltage and current. V01 and V02 are the waveforms of two output voltages on the DC side. U is the input voltage waveform on the AC side. It can be seen that the input current on the AC side has good sinusoidality, and a good phase tracking effect. The output voltage on the two support capacitors on the DC side remains basically the same, realizing the midpoint balance control.

4. Conclusion

In this paper, the working principle of a single-phase three-level ANPC rectifier and the working characteristics of power devices under three traditional SPWM modulation methods are analyzed. A modulation scheme based on two carrier cycles and two traditional SPWM1 is proposed. On this basis, the loss power model is established. The calculation shows that the alternating operation scheme of the two commutation modes can effectively reduce the loss unbalance on the premise of keeping the total loss basically unchanged.

References

[1] Tan G, Jing W. Active clamped three-level inverter and its junction temperature balance control[J]. Journal of Electrotechnology, 2012, 27(02):97-102.
[2] Zhang S, He W, Song W. Power Loss Equalization Method of Single-phase Three-level ANPC Rectifiers Based on Periodic Rotational Modulation[C]//2022 IEEE Transportation Electrification Conference and Expo, Asia-Pacific (ITEC Asia-Pacific). IEEE, 2022: 1-6.
[3] Luo l, Li Y, Li Z, et al. Loss equalization modulation method for switching devices of active neutral point clamped three-level inverter [J] New technology of electrical energy, 2021, 40 (04): 1-9.
[4] Jiao Y, Lee F C. New modulation scheme for three-level active neutral-point-clamped converter with loss and stress reduction[J]. IEEE Transactions on Industrial Electronics, 2015, 62(9): 5468-5479.
[5] Deng Y, Li J, Shin K H, et al. Improved modulation scheme for loss balancing of three-level active NPC converters[J]. IEEE Transactions on Power Electronics, 2016, 32(4): 2521-2532.
[6] Guo Y, Gao X, Tan J, et al. Research on loss balance modulation strategy based on single-phase half-bridge TL-ANPC inverter[C]//2022 5th International Conference on Energy, Electrical and Power Engineering (CEEPE). IEEE, 2022: 375-382.
[7] Zhang W, Zhang Z. SPWM-based loss-balancing control method for three-level ANPC inverters [J]. Power Electronics Technology, 2016, 50(09):4-6+12.
[8] Zhang B, Ge Q, Wang X, et al. Loss balance optimization control method for three-level active neutral point clamped converter [J] High voltage technology, 2016, 42 (09): 2775-84.
[9] Hu C, Ma D, Wang Q, et al. Loss distribution balance control strategy for three-level active neutral point clamped inverter [J] Journal of electrotechnics, 2017, 32 (01): 129-38.
[10] Jing X, He J, Demerdash N A O. Loss balancing SVPWM for active NPC converters[C]//2014 IEEE Applied Power Electronics Conference and Exposition-APEC 2014. IEEE, 2014: 281-288.

Workpieces detection algorithm based on improved YOLOv5

Jie Zhang [1, 2], **Tao Jiang** [1, 2*]

[1]School of Mechanical and Electrical Engineering, Changchun University of Science and Technology, Changchun 130022, China;

[2]Chongqing Research Institute, Changchun University of Science and Technology, Chongqing 401135, China

E-mail:1293798100@qq.com; E-mail: jiangtao@cust.edu.cn

Abstract: To solve the problems of similar product shapes, the low recognition rate of scattered workpieces and poor real-time performance in industrial sorting scenarios, a workpieces target detection algorithm based on improved YOLOv5 was proposed. The algorithm replaced the characteristic b network with a lighter ShuffleNetV2 network to achieve a lightweight network model, balance speed and accuracy. BiFPN structure was introduced to change the neck and enhance the ability of feature extraction at different scales. The research results indicate that the proposed improved algorithm has more advantages in industrial scenarios. It reduces the identification error rate of the target workpiece in the case of scattered and meets the requirements of accuracy and real-time recognition of the workpiece.

1. Introduction

Rapid and precise workpieces identification on the production line using visual technology has emerged as a research hotspot in intelligent manufacturing with the rapid progress of artificial intelligence technology [1]. On the production line, chaotic stacking of workpieces is still faced with great challenges in measurement and identification. Chaotic stacking of workpieces often blocks each other and interferes with each other, and its features are not obvious and have weak textures. Reflective scenes are frequent. A challenging issue is swiftly and effectively classifying the type of workpieces and creating the groundwork for the following capture.

Traditional object detection algorithms, such as the classical object detection algorithm based on the feature graph HOG [2] by Navneet Dalal and Bill Triggs, have low efficiency and poor robustness in workpiece detection. It is tough to meet the needs of industrial testing.

In latest years, deep learning has surpassed traditional methods in many fields [3]. It can also achieve good accuracy and efficiency in complex environments such as multi-target and low light. Such as R-CNN [4] and Faster R-CNN [5], where candidate regions need to be generated and then classified. YOLO is used for end-to-end detection [6]. At present, YOLOv5 is a relatively advanced target detection algorithm among current target detection algorithms [7]. Good results have also been achieved in various industrial problems [8].

To solve the trouble of recognition accuracy and speed of the traditional method, a target detection approach primarily based on YOLOv5 was proposed in this paper and improved. Firstly, the data set was made for the problem of no specific data set in the industrial scene. Then, the backbone feature network is replaced by the lighter ShuffleNetV2 network to achieve the lightweight, balanced speed and precision of the network model. The Neck structure was modified by combining the weighted bidirectional function pyramid community (BiFPN) structure, which enhanced the characteristic extraction and fusion process and improved the detection accuracy. Finally, experiments were conducted to verify that the workpieces can be quickly and accurately identified in the scattered mixed industrial scene.

2. Deep learning object detection algorithm

2.1 Network structure

YOLOV5 is a new member of the target detection family, and it adds a new strategy based on the previous one, which appreciably improves its detection pace and accuracy. The YOLOV5 model is small in size, quick in velocity and excessive in precision. The YOLOv5 structure is primary divided into four parts: input, backbone, neck, and output head. Input terminal mainly includes information enhancement, adaptive image filling, data preprocessing operation, etc. [9]. Backbone is a network used to extract features. It extracts information from images for later networks. The neck module splices different feature images together and sends them to the head of the detection head. Its algorithm structure is shown in Figure 1:

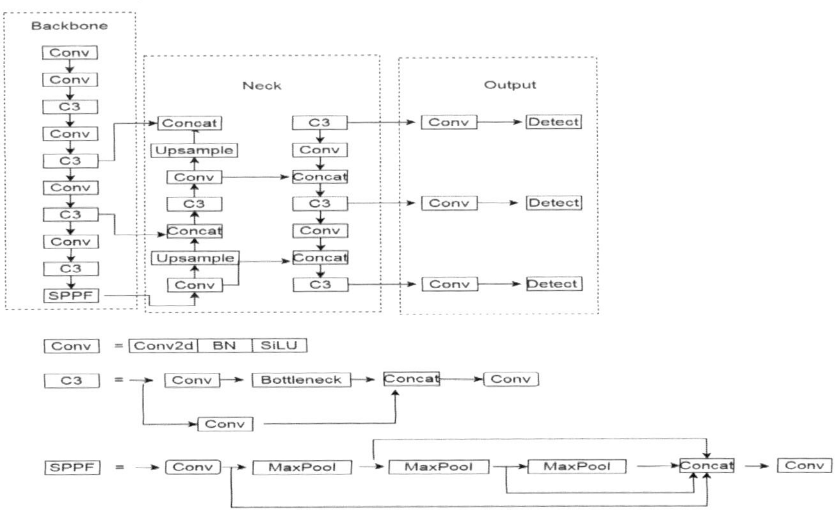

Figure 1. YOLOv5 algorithm structure diagram.

2.2 YOLOv5s algorithm improved

2.2.1 Replacement backbone network

The backbone feature network of Yolov5 adopts a c3 structure and has a giant wide variety of parameters, leading to slow detection speed and limited application. The study of small but efficient models will become more and more important. The improvement strategy is to try to exchange the backbone feature network with a lighter ShuffleNetV2 network to achieve lightweight, balance speed and precision of the network model [10]. Four design criteria for the lightweight convolutional neural network ShuffleNetV2 are proposed via a giant quantity of experiments. The influences of input/output channels, degree of network fragmentation, range of grouped convolutional groups, and element-by-element operation on speed and memory access on special hardware are analysed in detail. Channel Split is used to replace group convolution, which satisfies four design criteria and achieves the most suitable trade-off between velocity and precision. Its structure is shown in Figure 2.

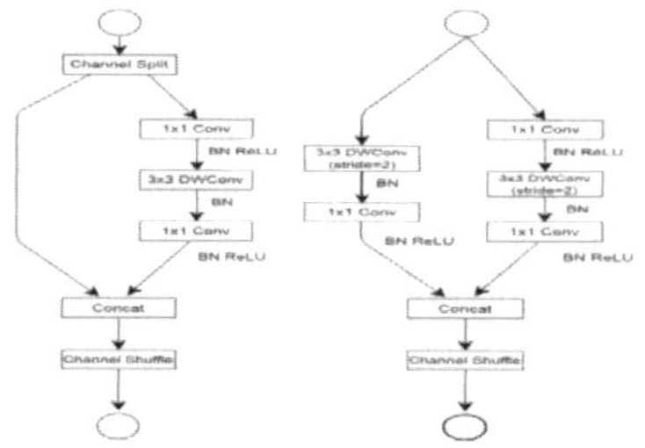

(a) Basic unit (b) Downsampling unit

Figure 2. ShuffleNetV2 network.

2.2.2 BiFPN structure is introduced to improve Neck

In the YOLOv5 network Neck structure, the proposal of PANet proves the effectiveness of bidirectional fusion, but its bidirectional fusion is relatively simple. On this basis, this paper tries to introduce the BiFPN structure to transform it, as shown in Figure 3:

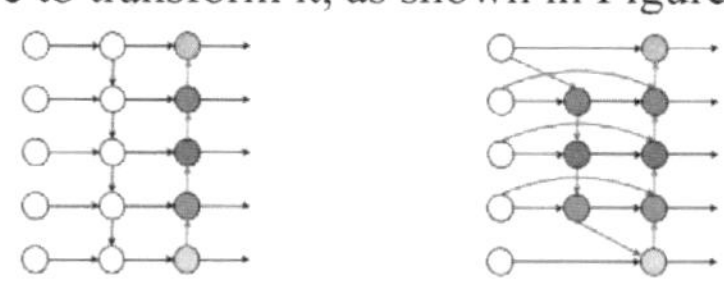

(a) PANeT (b) BiFPN

Figure 3. BiFPN structure diagram.

The BiFPN stands for Bidirectional Feature Pyramid Network weighted bidirectional feature pyramid network, an efficient bidirectional cross-scale weighted feature fusion method compared with PANet.

It mainly solves the following problems:

(1) Small target detection problem: BiFPN, as the feature network layer of repetitive modules, can obtain a higher level of feature fusion than PANet and increase the coupling of features of various scales, which is especially helpful for shallow features of small targets.

(2) Coincidence of workpieces: Because BiFPN adopts the cross-scale connection mode, the BIFPN enhances or restrains the expression of different detection features according to the cross-scale weight, to effectively alleviate the inaccurate identification caused by the coincidence of objects.

3. Experiment and result analysis

3.1 Data set construction

The self-made data set method was adopted to shoot and collect tee pipes, 500 images were taken from the core, and the workpieces to be identified were classified as 1. Tee pipe was labelled S, 2. Flange Core was labelled F. The data annotation interface is shown in Figure 4.

Figure 4. Data set annotation.

3.2 Experimental configuration

This experiment was carried out in the Windows 10 operating system. The program was written by Pycharm compiling environment, python language was used, and Pytorch was adopted as the deep learning framework. The network model adopted yolov5, the CUDA version was 11.1, and the generation number of model coaching was set to 100.

3.3 Experimental performance evaluation index

To make the results clearer, this paper verifies the performance of the improved algorithm from the aspects of accuracy, recall ratio and average accuracy. The calculation formulas are shown in Formulas (1), (2) and (3). The meanings of variables of evaluation indicators are shown in Table 1:

Table 1. Meanings of variables of evaluation indicators

Evaluation index	Positive	Negative
True	TP	TN
False	FP	FN

$$Precison = \frac{TP}{TP+FP} \quad (1)$$

$$Recall = \frac{TP}{TP+FN} \quad (2)$$

$$MAP = \frac{1}{Q}\sum_{q=1}^{Q} AP(q) \quad (3)$$

where Q is the number of classes.

3.4 Model performance analysis

After the training, the changes in each value tend to be stable. The performance indicators are shown in Figure 5.

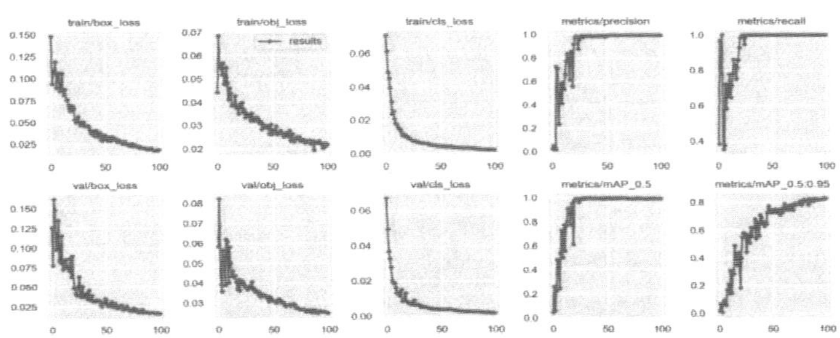

Figure 5. Various performance indicators.

3.5 Analysis of experimental results

In the experiment, two kinds of workpieces pictures are selected for recognition, and the recognition effect is output. The recognition effect of various workpieces under YOLOv5 and the improved YOLOv5 algorithm is shown in Figure 6:

Comparison before and after single target Tee pipe identification:

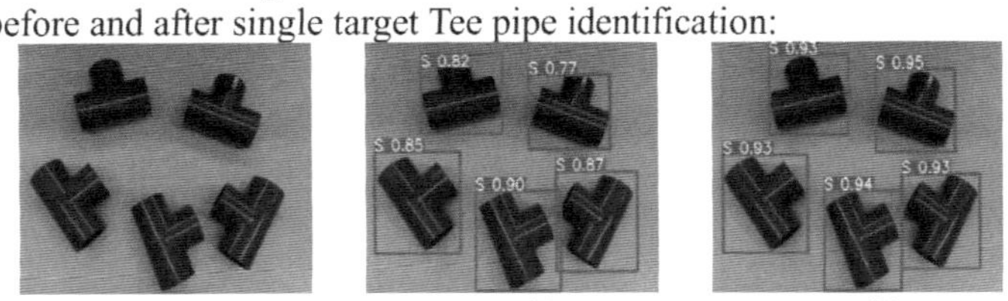

a (1) a (2) a (3)

Comparison before and after single target Flange Core identification:

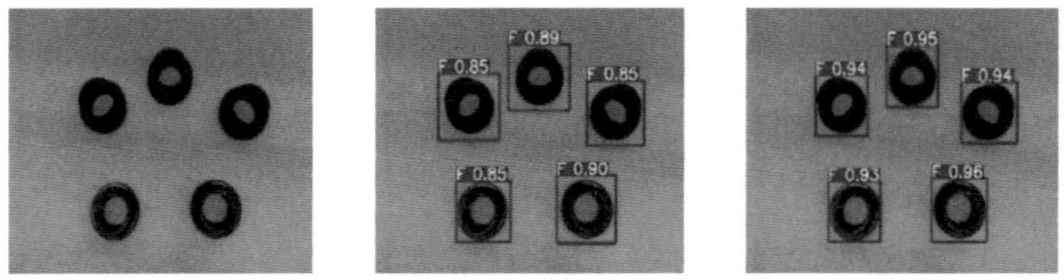

b (1) b (2) b (3)

Multi-target mixed recognition before and after comparison:

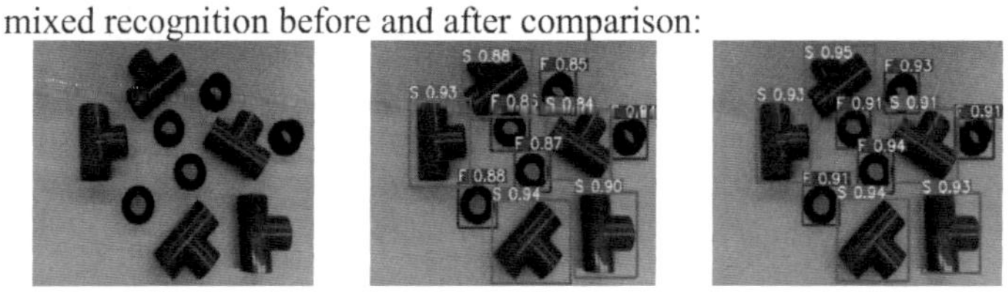

C (1) c (2) c (3)

Figure 6. Comparison of the effect of YOLOv5 and the improved YOLOv5 algorithm.

The traditional target detection algorithm, YOLOv5 algorithm and the algorithm in this paper were used to conduct a comparison experiment on the same set of data selected and record their recognition speed and accuracy, respectively. The accuracy and speed of workpiece identification are shown in Table 2.

Table 2. Effect verification and comparison

Algorithm	Category	Recognition accuracy	Recognition speed (amplitude/s)
Traditional algorithm	Tee pipe	0.788	1.051
	Flange Core	0.795	
YOLOv5s	Tee pipe	0.883	0.015
	Flange Core	0.895	
Algorithm of this paper	Tee pipe	0.932	0.016
	Flange Core	0.945	

From the analysis of the results of the three algorithms, it can be viewed that in contrast with the traditional target detection algorithm, the upgraded algorithm studied in this thesis has greatly elevated the recognition accuracy and recognition speed. Compared with the authentic YOLOv5 algorithm, the algorithm in this thesis has further extended the algorithm's accuracy. The recognition accuracy of tee pipe by the traditional algorithm is 78.8%, and that of Flange Core is 79.5%. The mean precision of tees using the YOLOv5 algorithm in this data set is 88.3%, the mean precision of Flange Core is 89.9%, the mean precision of tees using the improved algorithm in this data set is 93.2%, and the mean precision of Flange Core is 94.5%. Compared with the traditional algorithm, both speed and precision rate are appreciably improved. Compared with the original YOLOv5 algorithm, the recognition accuracy of the modified algorithm is increased by means of 5%.

According to the experimental results, in contrast with the common algorithm, the detection speed and accuracy of the upgraded algorithm studied in this thesis have been notably improved. Compared with YOLOv5, the improved algorithm of YOLOv5 can successfully identify the Tee pipe and the Flange Core in the industrial sorting scenario where the product shape is similar, and the workpieces are scattered. In the case of mixed occlusion, it can also successfully identify the workpieces to be grabbed with higher recognition accuracy.

4. Conclusion

In this paper, the upgraded yolov5 algorithm based totally on deep learning is studied to identify scattered workpieces. Experiments exhibit that the proposed algorithm is effective and stable and can meet the needs of industrial detection. The accuracy and speed of the improved algorithm have been greatly improved to meet the need for fast and accurate recognition of the workpiece. In subsequent studies, more types of workpieces identification studies will be considered to ensure model generalisation ability and recognition accuracy and speed while ensuring model diversity.

References

[1] Bao X L, Wang S Q. Survey of object detection algorithm based on deep learning[J]. Transducer and Microsystem Technologies,2022,41(04):5-9. DOI:10.13873/J.1000-9787(2022)04-0005-05.

[2] Dalal N. Histograms of oriented gradients for human detection[C]// Computer Vision and Pattern Recognition. IEEE Computer Society, 2005. DOI： 10.1109/cvpr.2005.177

[3] Lu H T, Zhang Q C. Applications of Deep Convolutional Neural Network in Computer Vision[J]. Journal of Data Acquisition and Processing,2016,31(01):1-17. DOI:10.16337/j.1004-9037.2016.01.001.

[4] Girshick R, Donahue J, Darrell T, et al. Rich Feature Hierarchies for Accurate Object Detection and Semantic Segmentation[J]. IEEE Computer Society, 2014. DOI: 10.1109/CVPR.2014.81

[5] Ren S, He K, Girshick R, et al. Faster R-CNN: Towards Real-Time Object Detection with Region Proposal Networks[J]. IEEE Transactions on Pattern Analysis & Machine Intelligence, 2017, 39(6):1137-1149. DOI:10.1109/TPAMI.2016.2577031

[6] Cao C, Zhang H Y. Face recognition algorithm based on improvedresidual network[J]. Transducer and Microsystem Technologies,2018,37(08):127-129+133. DOI:10.13873/j.1000-9787(2018)08-0127-03.

[7] Redmon J, Farhadi A. YOLO9000: Better, Faster, Stronger[J]. IEEE, 2017:6517-6525. DOI: 10.1109/CVPR.2017.690

[8] Ying Z, Lin Z, Wu Z, et al. A modified-YOLOv5s model for detection of wire braided hose defects[J]. Measurement, 2022(190-):190. DOI: 10.1016/J.MEASUREMENT.2021.110683

[9] Huang T Y, Hu B J, Zhu T T. Lightweight object detection method for road traffic scene[J]. Modern Electronic Technology,2022,45(03):88-95. DOI: 10.16652/j.issn.1004-373x.2022.03.018.

[10] Zhang X, Zhou X, Lin M, et al. ShuffleNet: An Extremely Efficient Convolutional Neural Network for Mobile Devices[J]. 2017. DOI:10.48550/arXiv.1707.01083

Numerical simulation of directional solidification process of single crystal blade based on precast

Xiaoyun Zhao [a], Guoqiang Cao [b*]

[1] College of Mechanical and Electrical Engineering, Shenyang Aerospace University, Shenyang, Liaoning, 110136, China

[a] E-mail: 985248733@qq.com

* Corresponding author: [b] e-mail: shenyangkong@126.com

Abstract: A finite element method was applied to establish a mathematical model of a single crystal blade's directional solidification process. The model had different angles between the axis of the mold group and the blade. The effects of different radiation angles on temperature variation laws at different blade positions were investigated. The temperature near the center was lower than that near the wall on the upper side of the baffle, while it was higher on the lower side. Moreover, the grain growth of stray particles in blade casting defects was influenced by blade orientation. The grain growth of scattering particles in blade casting defects was faster. The isolated solid-liquid mixing zone was more prone to form at the platform tip near the center when the blade orientation angle was smaller. The 0° orientation angle resulted in the blade's most severe stray grain phenomenon.

1. Introduction

A mathematical model based on precast software and the CAFE model was developed to simulate the directional solidification process of a single crystal superalloy blade for an aero-engine. This blade is critical but susceptible to casting defects like stray grains and large-angle grain boundaries during manufacturing [1]. Simulation can help improve the casting process and predict solidification defects [2, 3, 4], enabled by the rapid development of computer technology and computational materials science. Yang et al. [5] studied the mechanism of stray particle formation using a combined CA-FD model. They discovered that increasing shrinkage speed and inclination angle of isotherms would enhance undercooling in the platform area, facilitating stray particle formation and growth. Huo et al. [6] observed that a new region with large areas of dendrite fragments formed in the middle of the platform when the pull-out rate increased to 150 μm/s, impairing single crystal casting integrity. MA [7] revealed that grain defect formation was mainly associated with blade geometry and thermal conditions, such as thermal barriers at the platform end and edge plate. This obstructed single crystal growth continuation, causing grain branching or turning. Hence, studying blade orientation's relationship with aero-engine stray particles is vital for enhancing reliability. The simulation investigated how different blade orientations affected temperature distribution, solid-liquid mixing zone, secondary arm dendrite spacing, and grain impurity formation.

2. Physical model and simulation parameters

The directional solidification process under different blade orientations was numerically simulated by ProCAST software. The directional solidification equipment was somewhat simplified, as shown in Fig.

1, which included a heating zone, thermal baffle, cooling zone, and single crystal blade module. To improve the calculation speed, 1/4 entity is used for simulation analysis. Using ug11.0 to establish a three-dimensional solid model, the MeshCAST module is used for meshing, the two-dimensional meshing of the furnace's inner wall, and the casting's three-dimensional meshing. Fine mesh is used in important areas, such as blades, and coarse mesh is used in less important areas, such as the furnace body.

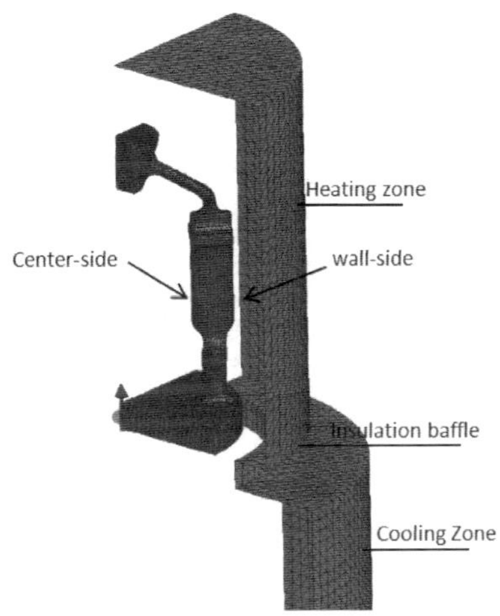

Figure 1. Simplified diagram of directional solidification equipment

As shown in Fig. 2, a model with different angles between the axial direction of the blade and the mold group is designed, and the blade is symmetrically distributed. Fig. 2 (a), (b), and (c) show the 0° component, 45° component, and 90° component, respectively.

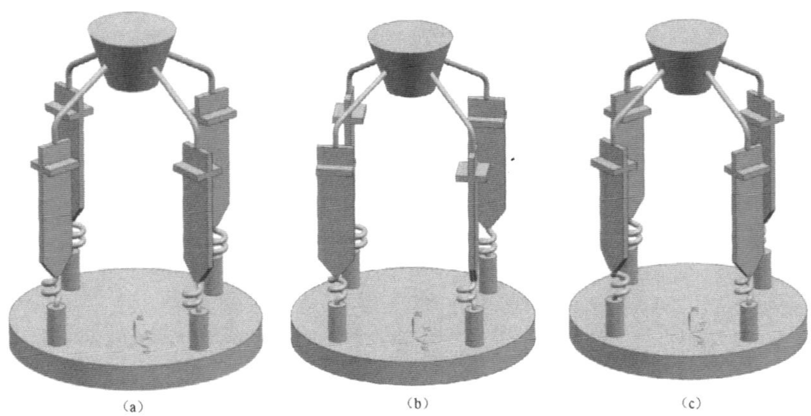

Figure 2. Different leaf orientation diagram, (a) 0°, (b) 45°, (c) 90°

This paper uses the second generation of single-crystal superalloy of nickel DD6. The pouring and heating zone temperatures are set to 1550 °C, and the drawing speed is 5 mm/min. The main parameters of DD6 superalloy simulation include liquidus temperature 1398 °C, solidus temperature 1343 °C, heat transfer coefficient of mold-water-cooled copper plate interface 100 $W \cdot m^{-2} \cdot °C^{-1}$, and heat transfer coefficient of casting-mold interface 500 $W \cdot m^{-2} \cdot °C^{-1}$ [8]. The casting-water-cooled copper plate interface heat transfer coefficient changes with temperature.

3. Effect of leaf orientation on temperature distribution and the solid-liquid mixing zone of the blade

The blade angle, which is the angle between the blade and the direction of the hot gas flow, affects the temperature distribution and the hot spot area of the blade. The temperature field is the temperature distribution on the blade surface, and the hot spot area is the region where the temperature exceeds a certain threshold. The blade angle influences the radiative heat transfer and the shielding effect between the blade and the furnace wall. Figure 3 shows the temperature distribution of the blade at t=1578 s under different blade orientations. When the blade angle is 0°, the blade is perpendicular to the gas flow, and the temperature distribution on the upper and lower sides of the blade is uneven. Due to the shielding effect, the cooling rate is slow, which leads to the temperature near the furnace wall lower than that of the center side, which is consistent with the findings of Yang et al. [9]. When the blade orientation angle is 45°, the temperature difference between the same surface of the blade and the furnace wall decreases, and the inclination of the isotherm also decreases. When the blade orientation angle is 90°, the blade surface is parallel to the furnace wall, and the points on the same surface of the blade body are equidistant from the furnace wall. The radiation heat transfer angle formed by the water-cooling ring is the same, and they have consistent heat transfer. Moreover, because the blade body is thin, the heat transfer effect of the blade body thickness can be ignored, and it has a more uniform temperature distribution.

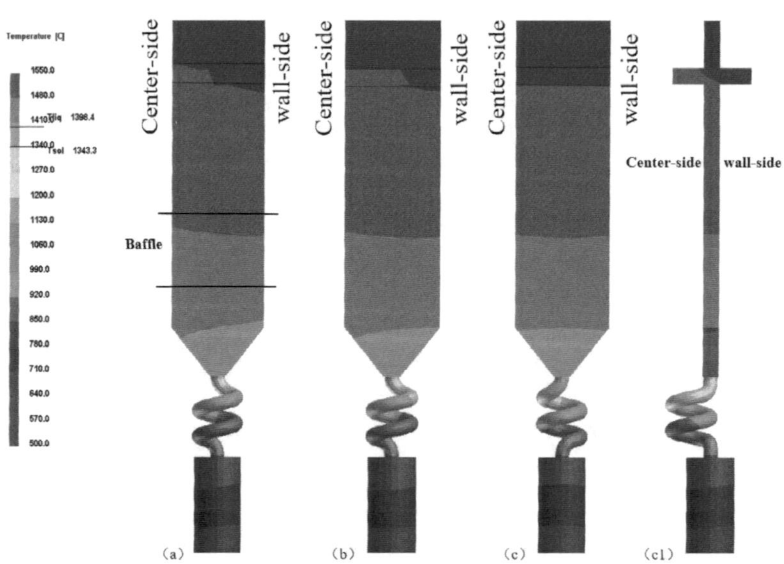

Figure 3. Temperature distribution under different blade orientations
(a) 0°, (b) 45°, (c) 90°and (c1) is a side view of (c)

Figure 4 shows the simulation results of the solid-liquid mixing zone reaching the platform position of the blade under different blade orientation angles when the withdrawal speed is 5 mm/min. When the blade angle is 0°, the solid-liquid mixing zone has a tilt angle, which makes the abrupt cross-section on the center side of the blade solidify earlier than the cross-section near the furnace wall. When the blade angle is 90°, the solid-liquid mixing zone at the tip of the platform has a certain tilt, forming an isolated hot spot area, which then develops into a stray crystal. Stray grains are a kind of crystal defect whose direction is different from the main crystal, which will reduce its mechanical properties. The shape and movement of the solid-liquid mixing zone depend on various factors, such as extraction speed, alloy composition, cooling rate, and thermal gradient. The exit velocity is 5 mm/min, and the blade angles are 0°, 45°, and 90°. These parameters influence the temperature and solute fields of the blade during directional solidification. A high extraction speed or a large blade angle increases the temperature gradient. It reduces solute enrichment in front of the solid-liquid boundary, which inhibits the formation of stray grains. A low extraction speed or a small blade angle can reduce the temperature gradient and increase the solute enrichment in front of the solid-liquid boundary, promoting spurious grain formation.

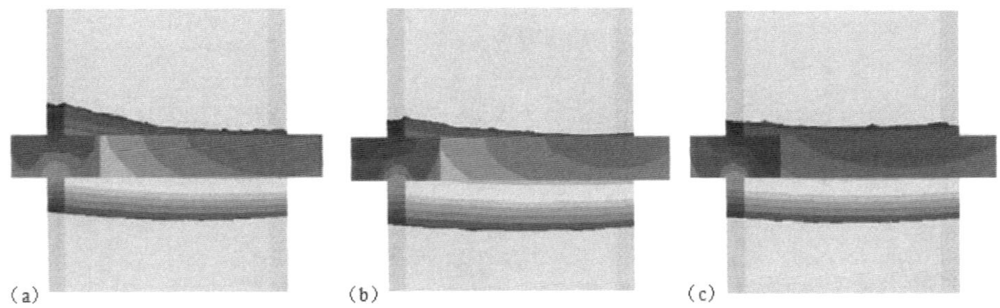

Figure 4. Solid-liquid mixing zone shape of blade platform position
(a) 0°, (b) 45°, (c) 90°

4. The effect of blade orientation on microstructure morphology

4.1. The effect of secondary arm dendrite spacing

As shown in Fig. 5, the secondary arm dendrite spacing ranges from 46 to 60 μm depending on the blade orientation, which agrees with the experimental results of Zhang et al. [10], demonstrating the accuracy of the calculation results. In addition, there is almost no difference in the secondary arm dendrite spacing of the induced crystal section under different orientations. This is because the water-cooled copper table strongly cools the induced crystal section, which is basically not affected by radiation heat transfer. As a result, the secondary arm dendrite spacing and blade orientation are unrelated. The middle and top portions of the blade body's secondary dendritic arm spacing are less spaced apart than the platform portion. This is due to the quicker cooling speed and bigger heat dissipation angle coefficient between the blade body and the cooling zone compared to the platform component. At the blade orientations of 45° and 0°, the secondary arm dendrite spacing is closer to the furnace wall side than the central side.

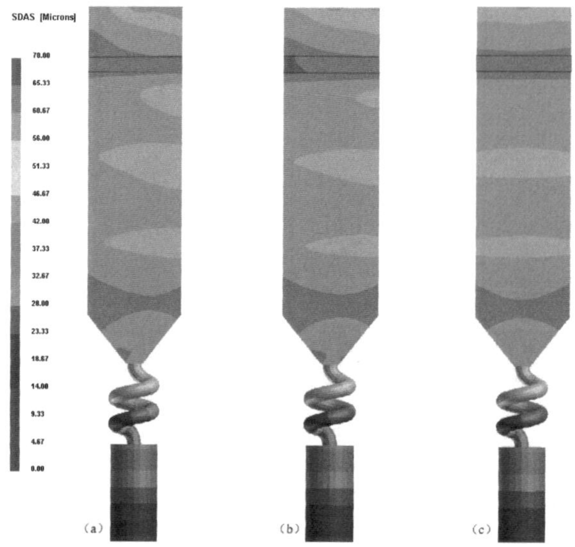

Figure 5. Secondary arm dendrite spacing under different leaf orientations
(a) 0°, (b) 45°, (c) 90°

4.2. Effect on the formation of stray grains

The basis for exploring the grain growth in the platform of the leaves was provided by analyzing temperature distribution and the solid-liquid mixing zone of the blade. The formation and growth of grains are affected by the changes in temperature distribution and the solid-liquid mixing zone of the blade., especially at the abrupt section of the platform. Therefore, different blade orientations were studied for their grain growth behavior. The grain growth simulation results under 0°, 45°, and 90° blade orientation were shown in Figure 6 at the 5 mm/min pulling speed. The results show that stray grains appear near the tip of the platform near the center side in both 90° and 45° directions, and the temperature distribution is gentle at the blade body. At the tip of the platform, there is a minimal effect from radiation heat transfer of the furnace wall. The temperature cooling rate is fast, and an isolated solid-liquid mixing zone is formed by tilting temperature distribution and then develops into stray grains. In the 0° direction of the blade sample, there is an increased tendency for stray crystal formation near the tip of the platform. This is because there is a tilt in temperature distribution and the solid-liquid mixing zone of the blade. Under this orientation and different temperature gradients at the same height, it favors stray grain formation, especially at abrupt sections of the blade. As the blade body position rotates perpendicular to the furnace wall, particle number and size increase significantly, and primary grain space decreases. Therefore, particle nucleation and growth tend to occur in the platform as the blade rotates toward the furnace wall's vertical position.

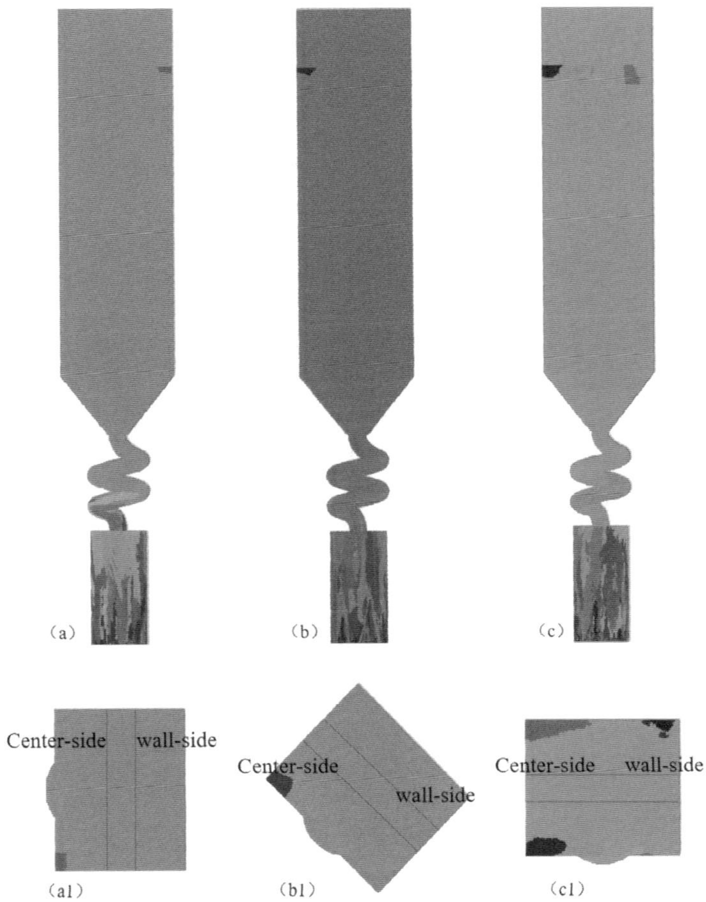

Figure 6. Grain structure under different blade orientations
(a) 0°, (b) 45°, (c) 90°

5. Conclusion

(1) Differently oriented blades' solid-liquid mixing zone and temperature distribution were simulated. The findings indicate that creating a solid-liquid mixing zone and a temperature distribution that is comparatively stable in the 90-degree orientated blade is simple. On the other hand, the blade's temperature distribution and solid-liquid mixing zone increasingly incline as the blade rotates perpendicular to the furnace wall.

(2) Due to the strong cooling effect of the water-cooled copper plate, there is little difference in the secondary arm dendrite spacing of the induced crystal section and the spiral section of different blade orientations. At the same height, the secondary arm dendrite spacing of the 90° oriented blade does not change significantly. In comparison, the secondary arm dendrite spacing of the center side of the 0° oriented blade is wider than that of the near wall side.

(3) When the blade position rotates perpendicular to the furnace wall, the number and size of the particles increase significantly, and the space occupied by the primary grains decreases, which is conducive to the nucleation and growth of stray particles in the platform.

References

[1] Ma, D X, Wang, F, Wu, Q. et al. (2017) Temperature evolution and grain defect formation during single crystal solidification of a blade cluster. China Foundry 14, 456-460.
[2] D M Liao, L Cao, F Sun, et al. (2018) Research Status and Prospect on Numerical Simulation Technology of Casting Macroscopic Process. Acta Metall Sin, 54(2): 161-173.
[3] Hu H J, Yang MB, Gong X B, et al. (2006) Numerical simulation optimization of casting process. Weapon Materials Science and Engineering, 6: 51-53.
[4] Lu Y Z, Shen J, Zheng W, et al. (2016) Numerical simulation of process optimization for solidification of single crystal castings[J]. Materials Engineering, 44(11): 1-8.
[5] Yang X L, Dong H B, Wang W, et al. Microscale Simulation of Stray Grain Formation in Investment Cast Turbine Blades[J]. Materials Science and Engineering A, 2004, 386: 129-139.
[6] M Huo, L Liu, W Yang, et al. (2019) Dendrite growth and defects formation with increasing withdrawal rates in the rejoined platforms of Ni-based single crystal superalloys, Vacuum, 161: 29-36.
[7] Ma D X. Novel casting processes for single-crystal turbine blades of superalloys. Frontiers of Mechanical Engineering [J]. 2018, 13(1): 3-16.
[8] F Qiu, K Bu, B Zheng, et al. (2020) Control of Edge Plate Stray Grain of Single-Crystal Turbine Blade by Using Process Bar Method, International Journal of Metalcasting, 14: 144-154.
[9] Yang L, Li J R, Jin H P, et al. (2014) Numerical simulation of directional solidification process of DD6 single crystal precision casting thin wall specimens. Materials Engineering, 11: 15-22.
[10] Zhang H, Xu Q Y, Shi Z X, et al. (2014) Numerical simulation study on directional solidification dendrite growth of DD6 high temperature alloy. Acta Metallurgica Sinica, 50(03): 345-354.

Parameter Optimization for Temporary Primary Frequency Control of Wind Turbines Based on Analytical Derivation

Xinyu Wei[1], Zhaoyang Jin[*1], Gengkun Li[1]

[1] Key Laboratory of Power System Intelligent Dispatch and Control of Ministry of Education Shandong University

zhaoyang.jin@sdu.edu.cn

Abstract. DFIG has the capability to participate in power system frequency modulation through inertial control by returning the energy stored in the rotating mass of the generator that is delivered to the grid. When frequency modulation occurs, the primary frequency drop can lead to a reduction in the rotor kinetic energy stored in the DFIG, which is subsequently used to produce extra active power to sustain the system frequency. However, once the available rotor kinetic energy is released, the output power of the DFIG abruptly drops to restore the rotor speed. To reduce the extent of the secondary frequency deviation caused by the inertial control of DFIGs, this paper applies an analytically derived approach to optimize the parameters of temporary primary frequency control during rotor speed recovery. Specifically, the parameters optimized include termination time and active power Simulation results conducted using DIGSILENT simulation software have demonstrated that the optimized temporary primary frequency control strategy is capable of effectively reducing the secondary frequency drop that may occur when the DFIG is removed from frequency control. Moreover, setting the optimal termination time and controlling the active power output while the rotor speed is being restored are crucial to prevent excessive mechanical wear and fatigue that may be induced by a too-rapid recovery of rotor speed.

1. Introduction

To address the environmental problem and energy crisis, the development of renewable energy, such as wind power, has been increasingly accelerated. However, unlike the traditional power systems whose frequency rarely fluctuates enormously [1]-[2], the new electrical grid is facing a range of problems caused by frequency fluctuations of large-scale new energy integration. Since they do not contribute to frequency control.

DFIG is widely recognized as a suitable model due to its exceptional features such as high-power generation efficiency and a small capacity frequency converter, as stated in previous studies. Unlike traditional synchronous generators, DFIG cannot directly regulate system frequency by releasing or storing rotor kinetic energy. Moreover, it typically operates in MPPT mode and lacks the standby capacity to participate in frequency control. As a result, scholars have started to show significant interest in the issue of system frequency dynamics [3].

To address the challenges mentioned above, a variety of methods have been proposed for wind turbines to participate in power system frequency modulation. These methods can be broadly divided into two categories: rotor kinetic energy control and load shedding control. While controlling the rotor kinetic energy can help the power system to regulate frequency, it can also result in secondary frequency drops when wind turbines exit from a frequency control state. Therefore, current research efforts are focused on minimizing the magnitude of these secondary frequency drops, and in some cases eliminating them altogether. In [4], an inertial controller for DFIG is proposed, which adjusts the output power based on RoCoF to address system frequency dynamics in wind turbines. Various other methods have been proposed in prior work to address system frequency dynamics in wind turbines. For instance, in [5], a droop controller for wind turbines was introduced, while [6] developed a frequency control method that combined inertial control with droop control. The concept of coordinated control between wind turbines and synchronous generators was introduced in [7], and [8] incorporated system frequency deviation and RoCoF into the wind turbine's active power control loop. Furthermore, [8] proposed a comprehensive inertial control approach to enhance the dynamic response of DFIG to the change in frequency. Additionally, [9] developed a variable parameter control method that employs integrated inertial control to regulate the release of rotor kinetic energy, thereby improving the minimum frequency drop and overall system frequency response.

Nowadays, the research of DFIG participating in frequency modulation has focused extensively on the improvement of methods, that is, to improve the secondary drop of frequency caused by wind turbine exiting frequency modulation by controlling the phase of wind turbine increasing output (overload phase) or the phase of rotor speed recovery after wind turbine exiting frequency modulation (speed recovery phase), but not much research has been done on the parameters of frequency modulation, such as incremental power, termination time, etc. To solve this problem, [10] gives the setting methods of the termination time of traditional inertia response respectively. Currently, the traditional inertia response approach of wind turbines has been largely phased out. However, existing methods such as those presented in [10] are not suitable for temporary primary frequency control. Thereby, in this paper, a parameter optimization method specifically for wind turbines employing temporary primary frequency control is proposed, and through simulation analysis, demonstrates that optimizing the control parameters can significantly reduce secondary frequency drops while making full use of the kinetic energy stored in the rotor of the wind turbine.

The structure of this paper is as follows: In Section 2, we present an introduction to temporary primary frequency control alongside traditional internal control. Sections 3 and 4 respectively introduce the calculation method of optimal termination time and active power in the rotor speed recovery phase. Section V introduces the simulation results. Section 6 summarizes this paper.

2. Temporary primary frequency control

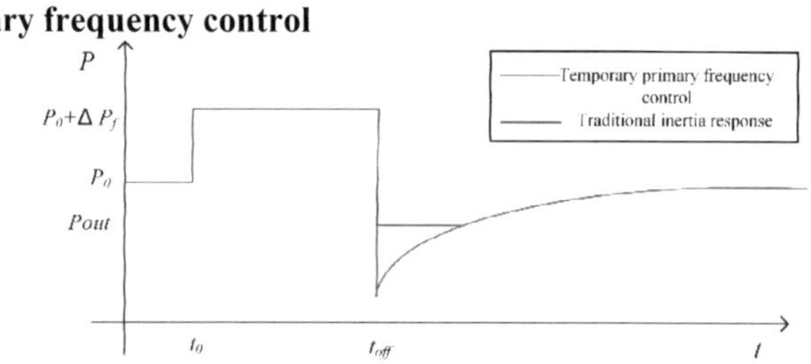

Figure 1. Temporary primary frequency control and inertia response

The schematic in Fig. 1 illustrates the traditional inertial response of DFIG. Specifically, in the event of active power disturbance, the DFIG increases its output by ΔP_f to help maintain the grid frequency. Upon exiting frequency control, the wind turbine's output rapidly drops to its MPPT operation point before the rotor speed is restored. These significant power drops can result in a large secondary frequency drop. To mitigate this secondary frequency drop by limiting the sudden drop in wind turbine output, a proposal is made in [11] to use temporary primary frequency control, represented by the red line in Fig. 1.

P_0 is the standard operation power of the DFIG, ΔP_f is the additional power of the DFIG, t_{off} is the termination time, t_0 is the time when the active disturbance event occurs, and P_{out} is the active power in the rotor speed recovery phase.

3. Calculation method of optimal termination time

3.1. SFR model

This study uses the SFR model to investigate the behavior of the system in response to changes in frequency. The SFR model constructs a simplified model by utilizing a simplified linkage model for the governor of a reheat steam turbine as an equivalent prime mover governor link. Due to its simplicity and constant order, regardless of the system size, the SFR model can be used to determine the maximum frequency deviation and the corresponding occurrence time under a specific disruption analytically. The schematic diagram representing the SFR model is depicted in Fig. 2. The SFR model's unit step response *h*(*t*), is expressed as:

$$h_{SFR}(t)=\frac{RP_{Step}}{DR+K_m}\left[1+ae^{-\varsigma\omega_n t}\sin(\omega_r t+\phi)\right] \tag{1}$$

$$a=\sqrt{\frac{1-2T_R\varsigma\omega_n+T_R^2\omega_n^2}{1-\varsigma^2}} \tag{2}$$

$$\omega_n^2=\frac{DR+K_m}{2HRT_R} \tag{3}$$

$$\varsigma=(\frac{2HR+(DR+K_mF_H)T_R}{2(DR+K_m)})\omega_n \tag{4}$$

$$\omega_r=\omega_n\sqrt{1-\varsigma^2} \tag{5}$$

$$\phi=\phi_1-\phi_2=\tan^{-1}(\frac{\omega_r T_R}{1-\varsigma\omega_n T_R})-\tan^{-1}(\frac{\sqrt{1-\varsigma^2}}{-\varsigma}) \tag{6}$$

Table 1 presents the quantity and SFR parameters of the power system used in this paper.

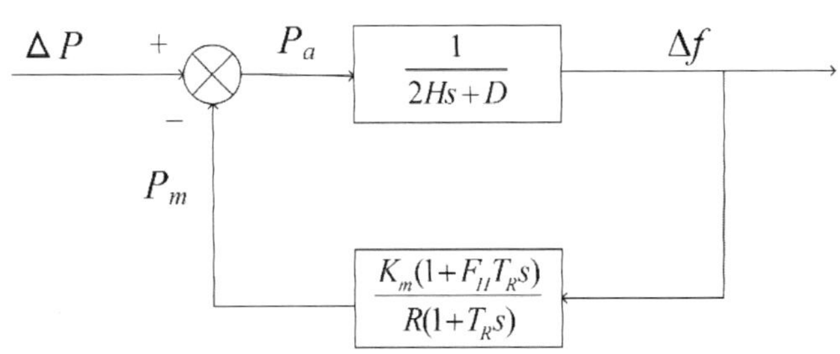

Figure 2. SFR model

Table 1. SFR PARAMETERS

Symbol	Quantity	Value
D	damping factor	0.01
R	regulation factor of the governor	0.05
T_R	reheat time constant	8(s)
K_m	mechanical power gain factor	0.95
F_H	faction of total power generated by the HP turbine	0.3
H	inertia constant	4(s)

3.2. Setting of parameter t_{off}

The equivalent load of wind turbines' temporary primary frequency control resulting from a disturbance is illustrated in Fig. 3. Accordingly, the dynamic process of the frequency response can be deemed as a composition of step responses at two different times, t_0 and t_{off}.

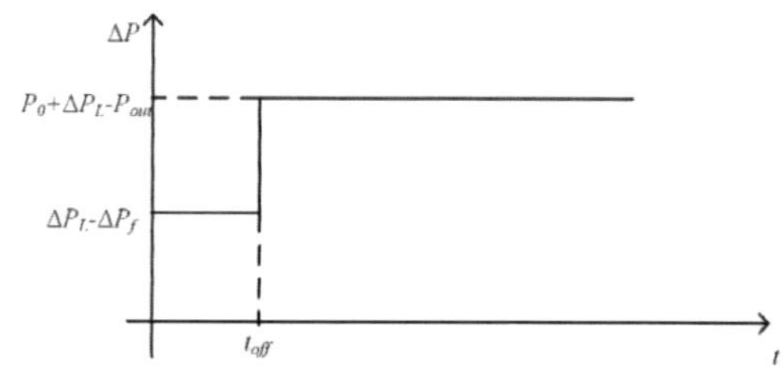

Figure 3. Load power change at synchronizer side

The total input power, ΔP, of the SFR model can be expressed as the summation of the following components:

$$\Delta P(t) = \Delta P_1(t) + \Delta P_2(t) \tag{7}$$

where $\Delta P_1(t)$ is caused by the event disturbance at t_0, which is supposed to be 0, $\Delta P_2(t)$ is caused by DFIG exit at t_{off}. Then the frequency response of each part can be expressed as:

$$\begin{aligned} \Delta f_1(t) &= (\Delta P_f - \Delta P_L)h(t)h_{SFR}(t) \\ \Delta f_2(t) &= (P_{\text{out}} - \Delta P_L - P_0)h(t - \text{t}_{\text{off}})h_{SFR}(t - \text{t}_{\text{off}}) \end{aligned} . \tag{8}$$

According to the superposition theorem:

$$\Delta f(t) = \Delta f_1(t) + \Delta f_2(t) . \tag{9}$$

By solving the partial derivatives of t and t_{off}, respectively, the optimal exit time toff can be obtained from the aforementioned equation in the following manner:

$$\begin{cases} \dfrac{\delta f(t, t_{off})}{\delta t} = 0 \\ \dfrac{\delta f(t, t_{off})}{\delta t_{off}} = 0 \end{cases} . \tag{10}$$

According to the response characteristics of SFR model, the time of DFIG exiting frequency modulation should be after the first frequency drop, otherwise, the frequency drop caused by wind turbine out of frequency modulation will be superimposed with the frequency drop caused by event disturbance, causing more serious frequency drop, which may start the low-frequency defense line of the system. According to the analysis of Equation (1), the time of the first frequency drop t_{n1} is:

$$t_{n1} = \frac{1}{\omega_r}\tan^{-1}\left(\frac{\omega_r T_R}{\zeta \omega_n T_R - 1}\right) \tag{11}$$

In order to minimize the secondary frequency drop that arises because of the wind turbine exiting frequency modulation after analyzing Equation (10), it is imperative to ensure that the termination time for the DFIG meets the following requirements:

$$t_{off} = t_{n1} + \left|\frac{\varphi_1}{\omega_r}\right| = \frac{1}{\omega_r}[\tan^{-1}\left(\frac{\omega_r T_R}{\zeta \omega_n T_R - 1}\right) + |\varphi_1|] . \tag{12}$$

4. Calculation method of optimal active power in rotor speed recovery phase

4.1. MPPT principle of DFIG

The mechanical power that is harnessed by a wind turbine, P_m can be expressed as follows:

$$P_m = \frac{1}{2}\rho A C_p(\lambda,\beta)V^3 \tag{13}$$

$$C_p(\lambda,\beta) = 0.2(\frac{116}{\lambda_i} - 0.4\beta - 5)e^{-\frac{12.5}{\lambda_i}} \tag{14}$$

$$\frac{1}{\lambda_i} = \frac{1}{\lambda + 0.08\beta} - \frac{0.035}{\beta^3 + 1} \tag{15}$$

$$\lambda = \frac{\omega_t R}{V} \tag{16}$$

where C_p (λ, β) is the wind energy utilization factor, λ is the tip speed ratio, V is the wind speed, β is the pitch angle, ω_t is the speed of the wind wheel, generator rotor speed ω_r=n*ω_t, n is the transformation ratio of gearbox.

According to Equations (14)-(16), when the wind speed is constant, the value of C_p is only related to the value of pitch angle and rotor speed. When the pitch angle is fixed, there is an optimal speed to where C_p obtain the maximum value. Furthermore, the maximum value of C_p, C_{pmax} is constant under different wind speeds. This constant is defined as the maximum wind energy conversion coefficient.

The aim of MPPT control is to facilitate the operation of a wind turbine at its optimum speed, thereby capturing the maximum available mechanical power. The operational point at which a wind turbine is capable of achieving the highest possible wind-energy conversion efficiency can be identified along the line that connects the maximum power point of the power-speed curve. By integrating C_{pmax} into Equation (13), and subsequently merging it with Equation 16, the resulting equation for the curve is:

$$P_{opt} = k_{opt}\omega_t^3 \tag{17}$$

$$k_{opt} = \frac{1}{2}\pi\rho\frac{C_{pmax}}{\lambda_{opt}^3}R^5 \tag{18}$$

where k_{opt} is the optimal power curve coefficient.

In light of Equation (17), the maximum power that DFIG can generate is directly proportional to the third power of the rotor speed. As k_{opt} remains constant, this establishes the optimal power curve unique to a given wind turbine.

4.2. Setting of parameter P_{out}

Illustrated in Fig. 4(a), red curve depict the rotor speed-power diagram of DFIG employing temporary primary frequency control, while the green curve represents the rotor speed-power diagram of the traditional inertial response.

The variables employed in this study are defined as follows:

ω_0 -rotor speed of the DFIG under normal operation
ω_{off} - rotor speed of the DFIG when exiting frequency control
P_{down} - power output of the wind turbine when the traditional inertia response exits frequency control.

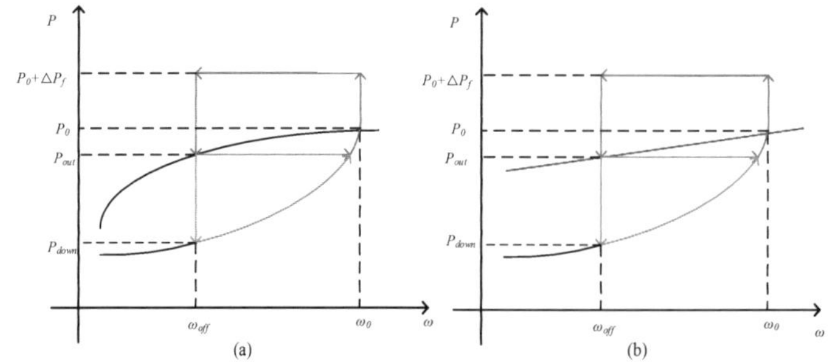

Figure 4. (a) Rotor speed power curve of wind turbine participating in frequency control (b) Simplified curve

In this study, we adopt the power platform as the intersection point between the wind turbine speed-power curve and mechanical power curve to ensure that DFIG's electromagnetic power remains below the mechanical power. This strategy enables smooth transition into the recovery stage, thereby minimizing the power drop associated with DFIG's exit from frequency control. Additionally, identifying the maximum power output of DFIG during the rotor speed recovery phase can effectively mitigate secondary frequency drops. Furthermore, we calculate the power drop attributable to both the temporary primary frequency control and traditional inertial response methods. The detailed procedure for determining P_{out} is presented below:

In the over-production stage, the electrical power output of the wind turbine is represented as $(P_0+\Delta P_f)$, while the mechanical power input can be determined using Equation (13). Accordingly, the following swing equation can be derived to describe the dynamics of wind turbines:

$$2H_{WT}\omega\frac{d\omega}{dt}=P_m(\omega)-(P_0+\Delta P_f) \tag{19}$$

where H_{WT} is the inertial constant of the wind turbine.

By integrating both sides of the equation over the entire over-production stage, we obtain the following expression:

$$\int_{\omega_0}^{\omega_{off}}\frac{2H_{WT}\omega}{P_m(\omega)-(P_0+\Delta P_f)}d\omega=\int_0^{t_{off}}dt=t_{off}\ . \tag{20}$$

Subsequently, we can utilize this equation to determine ω_{off}. Nonetheless, calculating the mechanical power input of the wind turbine constitutes a cumbersome process. Hence, we put forward an alternate approach to streamline the process, as demonstrated by the purple line in Fig. 4(b). In this method, the mechanical power input of the wind turbine can be simplified by a straight line passing through the origin. Consequently, Equation (13) can be simplified to:

$$P_m(\omega)=k_m\omega\ . \tag{21}$$

From the steady-state operating point:

$$P_m(\omega_0)=P_{opt}(\omega_0) \tag{22}$$

$$k_m=k_{opt}\omega_0^2 \tag{23}$$

where k_m is the simplified coefficient of mechanical power.

However, the differential equation shown in Equation (20) is still difficult to solve; the power disturbance ΔP_{off} when the wind turbine exits frequency modulation is approximated as a function of time.

Despite the presentation of the differential equation in Equation (20), obtaining a solution remains a challenging task. Therefore, we approximate the power disturbance ΔP_{off} that occurs when wind turbine exits frequency modulation as a function of time:

$$\Delta P_{off} = \Delta P_f + k_c(t_{off} - t_0) \tag{24}$$

where k_c can be solved by:

$$k_c = -\left.\frac{\mathrm{d}\Delta P_{off}}{\mathrm{d}t}\right|_{t=t_0} = -k_m \left.\frac{\mathrm{d}\omega}{\mathrm{d}t}\right|_{t=t_0} . \tag{25}$$

The rate of change of ω can be solved from Equation (19):

$$\left.\frac{\mathrm{d}\omega}{\mathrm{d}t}\right|_{t=t_0} = \left.\frac{P_m - P_e}{2H_{WT}\omega}\right|_{t=t_0} = \frac{P_0 - (P_0 + \Delta P_f)}{2H_{WT}\omega_0} = \frac{-\Delta P_f}{2H_{WT}\omega_0} . \tag{26}$$

So, the expression of k_c can be obtained by:

$$k_c = \frac{k_m \Delta P_f}{2H_{WT}\omega_0} = \frac{k_{opt}\omega_0 \Delta P_f}{2H_{WT}} . \tag{27}$$

At the same time, due to:

$$\Delta P_{off} = (P_0 + \Delta P_f) - k_m \omega_{off} \tag{28}$$

ω_{off} can be obtained by solving Equations (23), (24), (27) and (28):

$$\omega_{off} = \frac{P_0 - k_c t_{off}}{k_m} = \frac{P_0 - \dfrac{k_{opt}\omega_0 \Delta P_f}{2H_{WT}} t_{off}}{k_{opt}\omega_0^2} . \tag{29}$$

Therefore, P_{out}, P_{down} can be solved from the following equation:

$$P_{out} = k_m \omega_{off} = P_0 - \frac{k_{opt}\omega_0 \Delta P_f}{2H_{WT}} t_{off} \tag{30}$$

$$P_{down} = k_{opt}\omega_{off}^3 . \tag{31}$$

Then we can calculate the difference in the power drop ΔP_{change} between the temporary primary frequency control and traditional inertial control:

$$\Delta P_{change} = P_{out} - P_{down} = k_m \omega_{off} - k_{opt}\omega_{off}^3 . \tag{32}$$

4.3. Control strategy

The control logic for implementing temporary primary frequency control in the wind turbine, thereby facilitating its engagement in frequency control, is outlined as follows:

(1) We measure the real-time frequency deviation Δf and RoCoF using phasor measurement units (PMUs) at wind turbine; if either of the two parameters exceeds the limit, we start the frequency control of DFIG and proceed to step (2).
(2) We estimate the system frequency modulation parameters by the least square method based on the system state at the current time and determines the exit time t_{off}.
(3) We determine the value P_{out} of the power platform by Equation (30).

(4) We execute primary frequency control using to the given parameters.

5. Simulation results

5.1. Simulation settings

To evaluate the efficacy of the optimized temporary primary frequency control proposed in this study, we utilize the simulation model illustrated in Fig. 5. The DIgSILENT PowerFactory software is used to simulate and validate the control performance.

The model system utilized is comprised of a 50 MW wind power plant with a DFIG, along with three synchronous generators (SGs) having a collective capacity of 330 MVA (megavolt-ampere) and a static load of 200 MW. More specifically, the three SGs, rated for 150 MVA, 100 MVA, and 80 MVA capacity, respectively, are situated at the same bus within the three-bus model system. The governors for the SGs are modeled based on the IEEEG1 model.

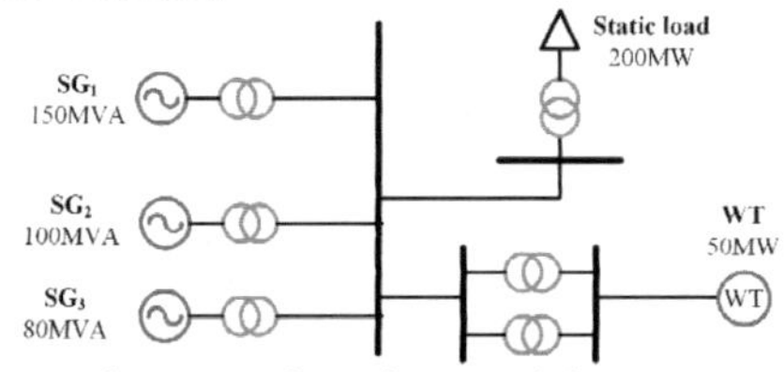

Figure 5. Three-bus model system

5.2. Simulation results

In this part, firstly, we use Equation (12) to solve t_{off}. t_{off} = 5 s can be obtained by least-square quasi combination calculation of system parameters. Then, select typical ΔP_f for simulation verification, respectively: (1) ΔP_f=0.08 p.u.; (2) ΔP_f=0.10 p.u. We can use Equation (31) to solve different value of P_{out} in different ΔP_f.

(1) Case 1: ΔP_f=0.08 p.u.

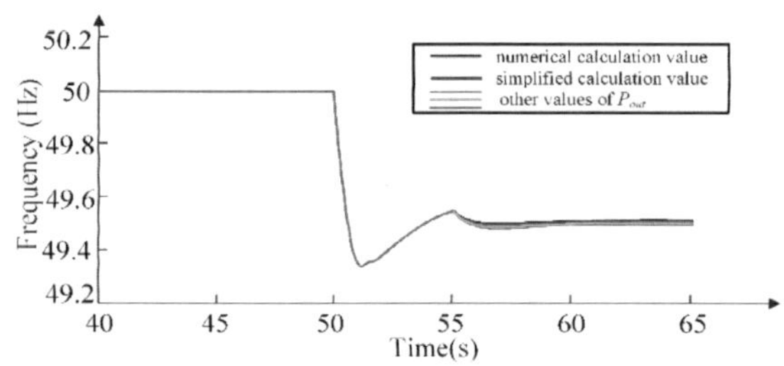

Figure 6. Frequency response of different values of P_{out}

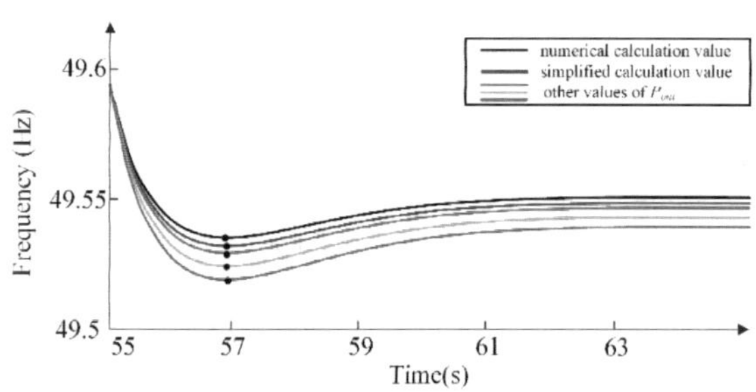

Figure 7. Secondary frequency drop of different values of P_{out}

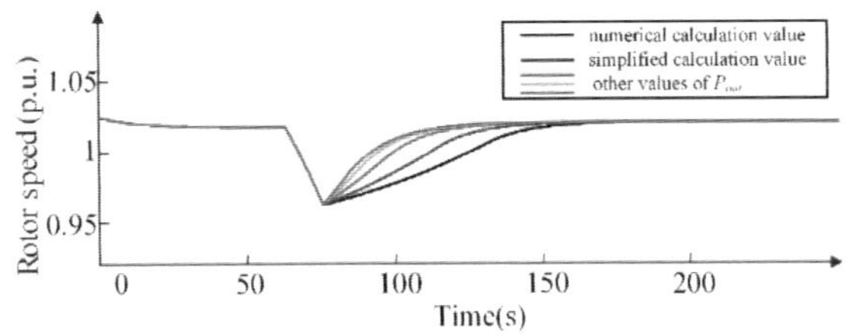

Figure 8. Rotor Speed of different values of P_{out}

(2) Case 2: ΔP_f =0.10 p.u.

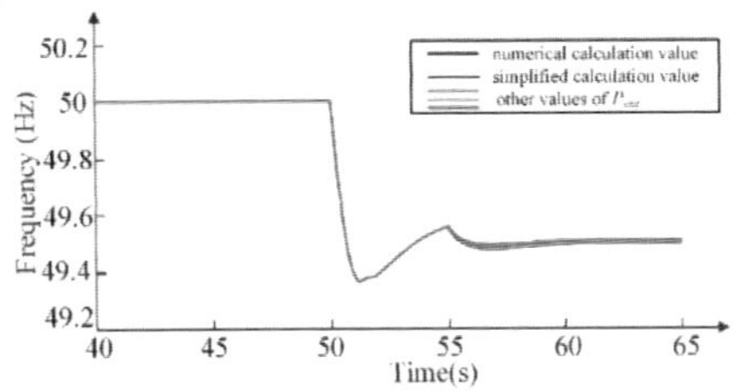

Figure 9. Frequency response of different values of P_{out}

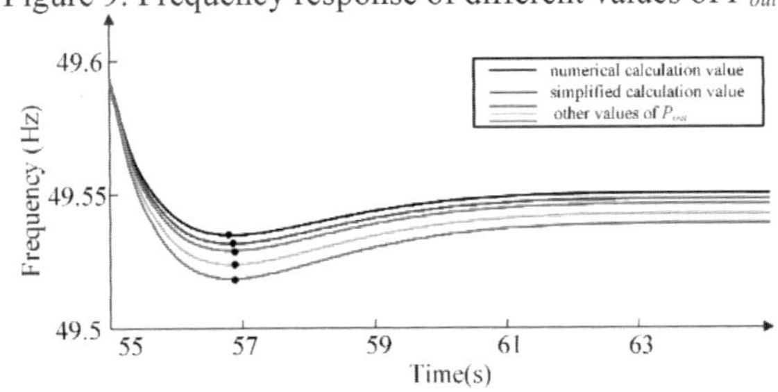

Figure 10. Secondary frequency drop of different values of P_{out}

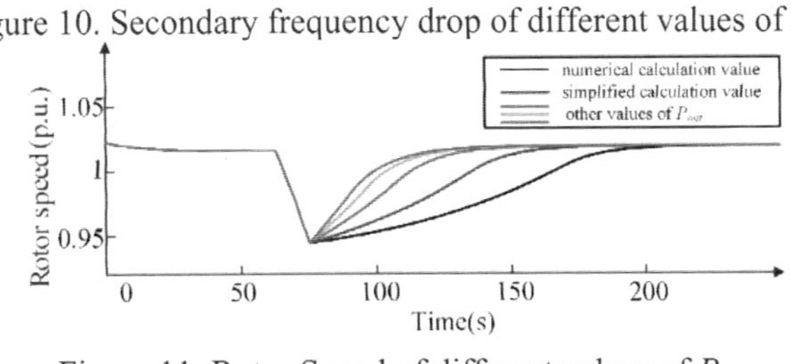

Figure 11. Rotor Speed of different values of P_{out}

Figs. 6-11 present the frequency response and rotor speed for the exact numerical calculation of the accurate P_{out} value, represented by the black curve, as well as P_{out} obtained using our simplified calculation method, depicted by the blue curve. For the two simulation examples carried out, ΔP_f and t_{off} are fixed. During the power support stage of DFIG, specifically, the period of initial frequency drop, it is observed that the various P_{out} values, irrespective of their specific magnitude, exhibit a common minimum frequency and rotor speed nadir. The primary difference lies in the lowest point of the secondary frequency drop and rotor speed recovery process. Additionally, Fig. 6 and Fig. 9 illustrate the frequency response under event disturbance. As demonstrated in Fig. 7 and Fig. 10, the simplified calculation of P_{out} results in a higher minimum frequency point than that obtained with other values of P_{out}. Notably, raising the lowest point of the secondary frequency drop elevates the steady-state frequency, and the variation between the simplified calculation and numerical calculation results is minimal. Compared to the numerical calculation, the simplified calculation utilized in this study delivers the P_{out} value suitable for the wind turbine more rapidly, enabling the wind turbine to engage in frequency modulation. As depicted in Fig. 8 and Fig. 11, the numerical calculation results in a slow rotor speed recovery process. The optimization method proposed in this study can expedite the rotor speed recovery while achieving greater system frequency support. Compared to traditional inertial response, the slower rotor speed recovery can mitigate the mechanical fatigue issue resulting from the overly rapid recovery of wind turbine speeds. Therefore, the method recommended in this work is technically and methodologically viable.

6. Conclusion

This paper presents an optimization design for the parameters of DFIG based on temporary primary frequency control. Our simplified analysis yields a direct method for computing the power platform value. The results obtained from simulations, using the model system detailed in Section IV, illustrate that our proposed optimization method delivers the same once frequency nadir as well as the following advantages when compared to other values of P_{out}:

(1) The derivation process of P_{out} and t_{off} is given to make full use of the frequency modulation ability of the wind turbine.

(2) The simplified calculation value has a higher frequency secondary drop lowest point than other values, and the effect is similar to the numerical calculation value.

(3) The simplified calculation method slows down the rotor speed recovery process and avoids the mechanical fatigue caused by too fast rotor speed recovery.

One limitation of the parameter optimization method presented in this study is that the optimization algorithm for ΔP_f is not tailored to current wind conditions. Future research efforts will be aimed at addressing this issue.

References

[1] J Zeng and J Huang, “Dynamic Frequency Control of Power Systems Based on State Feedback Linearization,” Grid Analysis&Study, vol. 45, no. 9, pp. 40-44, China. Sept 2017.

[2] Y Wu, B Yan, et al., “Coordinated Frequency and Voltage Control Strategy for Microgrid,” Shanxi Electric Power, vol.45, no.2, pp. 15-20, China. Feb 2017

[3] J Ren, “Discussion on strategy formulation of regional power grid security and stability control system,” Electrical technology, pp.13-15, China. Oct 2016.

[4] J. Ekanayake and N. Jenkins, “Comparison of the response of doubly-fed and fixed-speed induction generator wind turbines to changes in network frequency,” in IEEE Transactions on Energy Conversion, vol. 19, no. 4, pp. 800-802, Dec. 2004.

[5] R. G. de Almeida and J. A. Pecas Lopes, “Participation of Doubly Fed Induction Wind Generators in System Frequency Regulation,” in IEEE Transactions on Power Systems, vol. 22, no. 3, pp. 944-950, Aug. 2007.

[6] J. Morren, et al. “Wind turbines emulating inertia and supporting primary frequency control,” in IEEE Transactions on Power Systems, vol. 21, no. 1, pp. 433-434, Feb. 2006.

[7] J Cao, H Wang, J Qiu. “Frequency Control Strategy of Variable-speed Constant-frequency Doubly-fed Induction Generator Wind Turbines,” Automation of Electric Power Systems, vol. 33, no. 13, pp.78-82, Jul. 2009, DOI: 10.7500/AEPS200903098.

[8] X Qiu, Z Wu, B Chen, X Zhang, G Wang. “Improved Frequency Control Strategy for Variable-speed Constant-frequency Double-fed Induction Generator Wind Turbines,” Proceeding of the CSU-EPSA, vol. 23, no. 3, pp. 63-68, Jun. 2011

[9] R YOU, et al. “Frequency support capability of variable speed wind turbine based on the electromagnetic coupler,” Renewable Energy, vol.74, pp.681-688, Feb 2015

[10] Y Guo, W Bao, et al. “Analytically derived fixed termination time for stepwise inertial control of wind turbines” International Journal of Electrical Power & Energy Systems, 2020, 121:106106.

[11] R. T. Byerly et al., “Dynamic models for steam and hydro turbines in power system studies,” IEEE Trans. Power Apparatus. Syst., vol. PAS-92, no. 6, pp. 1904–1915, Nov. 1973.

Intelligent Parameter Identification and Sensorless Control of Permanent Magnet Synchronous Motor

Tongyue Chen[1], **Tiantian Liang**[2, *]

[1,2] School of Automation and Electrical Engineering, Dalian Jiaotong University, Shahekou, Dalian, Liaoning, China

*Corresponding author: liangtiantian1122@163.com

Abstract. In this paper, a spider monkey optimisation (SMO) algorithm is utilised to identify the parameters of the permanent magnet synchronous motor (PMSM), considering the parameters vary during the motor operation, which affects the sensorless control (SC) performance of the motor. An improved sliding mode observer (SOBS) is proposed for estimating the position and speed of the rotor. First, the SMO algorithm is used to identify the parameters of PMSM. Then, based on the identification results, an improved SOBS is proposed by a piecewise Sigmoid function. Furthermore, the stator position and speed are estimated by extended state observer (ESO) and phase-locked loop (PLL). Finally, a comparison simulation scenario is provided to demonstrate the efficacy of the suggested approach.

1. Introduction

Currently, the sensorless control technology of PMSM has two control methods: the high-frequency signal injection method and the method of a mathematical model based on the fundamental wave [1]. The method based on the fundamental wave mathematical model is mainly applied to medium or high-speed motors, such as the model reference adaptive method [2], extended Kalman filter method, and SOBS method. Among these, the SOBS has attracted much attention because it can implement nonlinear control, can have strong robustness and can have simple calculations. Researchers proposed the inverse tangent function to extract the angle and speed information of the inverse electromotive force [3]. However, using the inverse tangent function, the noise is directly introduced into the inverse tangent calculation. Constructing a back electromotive force observer has been proposed to replace the arctangent function [4]. However, the sliding mode observer-based method increases the amount of computation. Furthermore, the methods above are easily affected by the variation of motor parameters [5].

Some scholars have studied the problem by combining parameter identification with a sensorless algorithm [6,7]. The RLS algorithm has been combined with extended back electromotive force and has been used to identify motor parameters [8]. Then, based on the identification results, the position sensorless control was implemented. However, the RLS algorithm requires a large amount of computation. Chen and Chen identified PMSM resistors and inductance [9]. They input the identification value to a low-pass filter for filtering so that the influence of parameter changes on sensorless control is decreased. However, the parameters in these studies are hard to update in real-time.

Based on these reasons, this paper proposes a spider monkey intelligent algorithm for online identification of the stator resistance, *d-q* axis inductance and rotor flux of PMSM. To estimate the rotor position and rotor speed of PMSM, the SOBS is used in conjunction with PLL and ESO. The innovations of this paper are as follows:

1) The use of the SMO algorithm in the SOBS improves the vector control performance.

2) The switching function in the SOBS is replaced by the segmental Sigmoid function so that the high-frequency chattering can be suppressed.

3) Rotor position and velocity are estimated using the ESO-PLL, which performs well in estimating position and speed in the low-speed range.

2. PMSM parameter identification based on the spider monkey algorithm

2.1 Mathematical model of PMSM

In this paper, the mathematical model of PMSM in the rotational coordinate system of *d-q* axis is adopted as the object of study, and the core saturation, eddy current loss and iron loss are ignored. Thus, the state-space expression of PMSM is expressed as Formula (1):

$$\begin{bmatrix} v_d \\ v_q \end{bmatrix} = \begin{bmatrix} R_s & -\omega_r L_q \\ \omega_r l_d & R_s \end{bmatrix} \begin{bmatrix} i_d \\ i_q \end{bmatrix} + \begin{bmatrix} 0 \\ \omega_r \psi_f \end{bmatrix} \tag{1}$$

where i_x, v_x, L_x are the current component, voltage component, and inductance on the *d*-axis or *q*-axis, and R_s, ψ_f and ω_r denote the stator resistance, the magnetic flux and the electric angular velocity. Moreover, the vector control method of $i_d = 0$ is adopted to control PMSM and injected a non-zero current into the *d*-axis to obtain another two discretised voltage formulas. Then, the PMSM parameter identification model is established as Formula (2):

$$\begin{cases} v_{d0}'(t) = -L_q \omega_{r0}(t) i_{q0}(t) \\ v_{q0}'(t) = R_s i_{q0}(t) + \psi_f \omega_{r0}(t) \\ v_d'(t) = R_s i_d(t) - L_q \omega_r(t) i_q(t) \\ v_q'(t) = R_s i_q(t) + L_d \omega_r(t) i_d(t) + \psi_f \omega_r(t) \end{cases} \tag{2}$$

where $v_x'(t)$ is the identification values at *t* time, and $i_x(t)$ and $\omega_x(t)$ are the actual values at *t* time.

2.2 Spider monkey optimisation

In the SMO, each individual represents a solution; the optimal solution is found by exploring the entire region[10]. The SMO consists of the following steps:

1. Initialization phase: Formula (3).

$$E_{ij} = E_{\min j} + U(0,1) \times (E_{\max j} - E_{\min j}) \tag{3}$$

where E_i represents the *ith* individual, $E_{\min j}$ and $E_{\max j}$ are the maximum thresholds of dimension *j*, and $U(0,1)$ are random numbers distributed within (0,1).

2. Local leader stage: Update according to Formula (4).

$$E_{newij} = E_{ij} + U(0,1) \times (LL_{kj} - E_{ij}) + U(-1,1) \times (E_{rj} - E_{ij}) \tag{4}$$

where E_{newij} is the updated value, E_{ij} is the *ith* individual in dimension *j*, LL_{kj} is a member of group *k* in dimension *j*, E_{rj} is the spider monkey of other groups except the group *i* in dimension *j*.

3. World leader stage: Formula (5) is the fitness function of the *ith* spider monkey, where F_i is the objective function; Formula (6) is the *ith* spider for the fitness probability function monkey, which means the probability of this spider monkey being selected as the world leader (WL).

$$fit_i = \begin{cases} 1+|F_i|, F_i < 0 \\ 1/1+F_i, F_i \geq 0 \end{cases} \tag{5}$$

$$prob_i = 0.9\left(fit_i / \max\_fit\right) + 0.1 \tag{6}$$

The number randomly generated between (0,1) is compared with $prob_i$. If $prob_i$ is larger, a random dimension of the *ith* spider monkey is selected for updating; otherwise, it is not updated. The updated formula is given as Formula (7):

$$E_{newij} = E_{ij} + U(0,1) \times (WL_j - E_{ij}) + U(-1,1) \times (E_{rj} - E_{ij}) \tag{7}$$

where WL_j is the WL on the *jth* dimension.

4. World leader learning stage: Greedy selection is made according to the fitness function calculated in the WLS stage for selecting the best spider monkey in the group to be the world leader. Meanwhile, World Limit Count (WLC) records the number of world leaders updating.

5. Local leader learning stage: The fitness function is greedily selected within each group to find the optimal solution and become the local leader. Meanwhile, each group has a Local Limit Count (LLC). A maximum value is set for the LLC. When the counter reaches this maximum value, the algorithm is considered to trap into a locally optimal solution.

6. Local leader decision stage: If the LLC increases by 1, all spider monkeys in the group update based on the positions of the local leaders according to Formula (8):

$$E_{newij} = E_{ij} + U(0,1) \times (WL_j - E_{ij}) + U(0,1) \times (E_{ij} - LL_{ij}) \tag{8}$$

If LLC is not increased, this phase is skipped.

7. World leader decision stage: In this stage, the maximum value of the WLC is set. When the WLC exceeds the maximum value, the spider monkey population divides into smaller subgroups or re-fusion groups. The molecular grouping rule is that a small group of monkeys cannot exceed the number of $N/10$.

2.3 Parameter identification of PMSM based on SMO

PMSM parameter identification can be regarded as finding the optimal solution. The difference between the actual value and the output value of the identification model established in this paper constitutes the fitness function. The optimal solution is found by the SMO in the optimisation process. The value of $f(\theta)$ is reduced by iterating. A smaller fitness function value ($f(\theta)$) indicates a better identification value. The fitness function is as Formula (9):

$$f(\theta) = \left|v_{d0}(t) - v_{d0}'(t)\right| + \left|v_{q0}(t) - v_{q0}'(t)\right| + \left|v_{d1}(t) - v_{d1}'(t)\right| + \left|v_{q1}(t) - v_{q1}'(t)\right| \tag{9}$$

where v_{d0}, v_{q0} v_{d1} and v_{q1} are the actual values collected, v_{d0}', v_{q0}', v_{d1}' and v_{q1}' are the calculated values by the SMO. Figure 1 is the identification principal diagram.

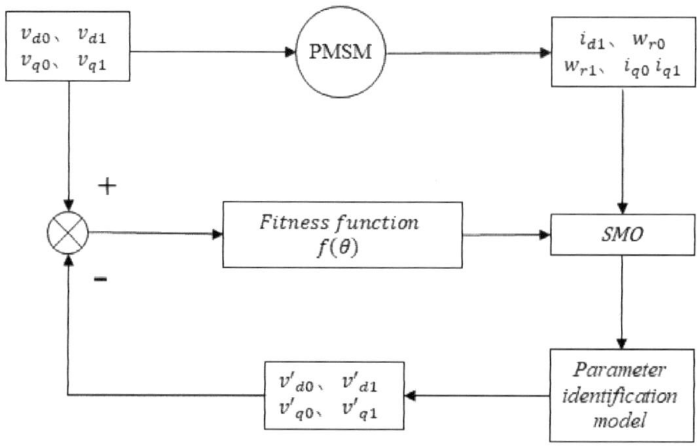

Figure 1. Flow chart of the SMO identification

3. Speed sensorless control based on improved SOBS

3.1 Improved SOBS of PMSM

This paper proposes a three-phase Sig function to replace the switching function. The three-phase Sig function contains two boundary layer coefficients, η and α . η is similar to the cutoff coefficient, and the function value between $|\eta|$ is 0. The smaller η is, the poorer the robustness and the stronger the chattering inhibition ability is. α determines the slope of the curve. The larger α is, the less robust and the faster the system responds. Otherwise, the stronger the robustness is, the slower the system response will be. The adjustment of η and α is more robust while reducing system chattering. Formula (10) is the segmental Sig function.

$$S(x)=\begin{cases} sig(x-\eta) & ,x\geq\eta \\ 0 & ,|x|=\eta \\ sig(x+\eta) & ,x\leq\eta \end{cases} \tag{10}$$

When PMSM runs at medium and low speeds, the impact of stator resistance variation on rotor position estimation can be expressed as Formula (11):

$$\theta=-\arctan\left[\left(-R_s i_\alpha+v_\alpha-L_s\frac{di_\alpha}{dt}\right)\Big/\left(-R_s i_\beta+v_\beta-L_s\frac{di_\beta}{dt}\right)\right] \tag{11}$$

According to the above section, the SMO algorithm is used for the on-line identification of PMSM resistance. The identification results are input into the current SOBS to improve the estimation accuracy of the SOBS.

Combined with the above two points, the improved SOBS reduces the chattering problem caused by the constant switching of switching functions and decreases the influence of motor parameter changes on sensorless control. Figure 2 is the improved sliding mode structure diagram.

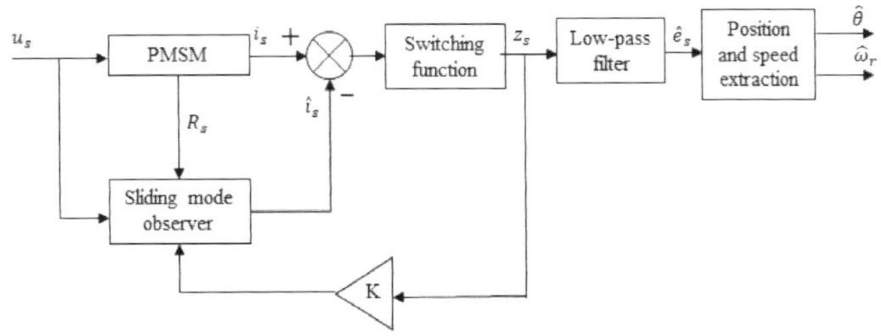

Figure 2. Improved sliding mode structure diagram

The formula is given as Formula (12).

$$\begin{cases} L_s \dfrac{d\hat{i}_\alpha}{dt} = -\hat{R}_s \hat{i}_\alpha + v_\alpha - gS\left(\hat{i}_\alpha - i_\alpha\right) \\ L_s \dfrac{d\hat{i}_\beta}{dt} = -\hat{R}_s \hat{i}_\beta + v_\beta - gS\left(\hat{i}_\beta - i_\beta\right) \end{cases} \tag{12}$$

where $S(x)$ is the segmented Sigmoid function, $\hat{r}_s$ is the stator resistance identified online.

3.2 Rotor position and speed estimation based on ESO-PLL

PLL consists of a phase discriminator (PD), a loop filter (LF) and a voltage-controlled oscillator (VCO). Differential operation is avoided when PI adjustment is carried out in the LF link, and the rotational speed is calculated in a closed-loop form. Since the differential of the electric angle in PMSM is the speed, and the differential of the speed is the angular acceleration, this paper adopts the ESO and the PLL to track the speed of PMSM. The electric angle and speed are taken as the state variables of the observer, and the acceleration is taken as the state variables of the ESO. In the ESO-PLL, the rotor position $\hat{\theta}_e$ is calculated according to the back electromotive force $\hat{e}_s$ output by low-pass filtering. The rotor position $\hat{\theta}$ estimated by the PLL and $\hat{\theta}_e$ is subtracted as the PD link. ESO is given as an LF link, and the angle estimation in the ESO is given as a VCO link. Figure 3 is the structure diagram of the ESO-PLL.

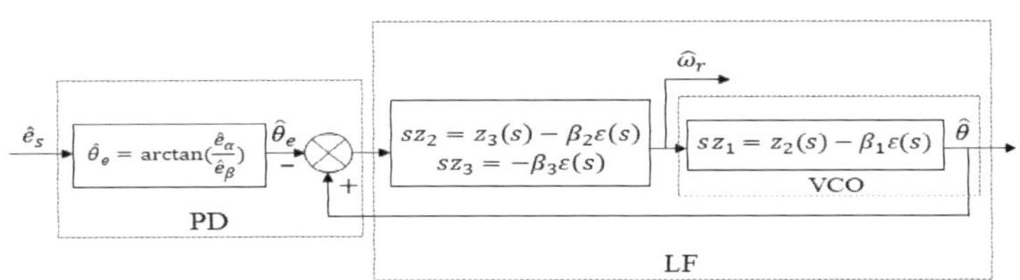

Figure 3. The structure diagram of the ESO-PLL

The closed-loop transfer function and error transfer function of ESO-PLL are in Formula (13).

$$\begin{cases} G(s) = \dfrac{\hat{\theta}}{\hat{\theta}_e} = \dfrac{\beta_1 s^2 + \beta_2 s + \beta_3}{s^3 + \beta_1 s^2 + \beta_2 s + \beta_3} \\ G_e(s) = \dfrac{e_2(s)}{d(s)} = \dfrac{-(s+\beta_1)}{s^3 + \beta_1 s^2 + \beta_2 s + \beta_3} \end{cases} \tag{13}$$

If $d(s) = C/s$, we multiply both sides of the error transfer function by $d(s)$, and find the steady-state error e_2 by the final value theorem (Formula 14):

$$\lim_{t\to\infty}|e_2| = \left|\lim_{s\to\infty} s\frac{C}{s}\frac{-(s+\beta_1)}{s^3+\beta_1 s^2+\beta_2 s+\beta_3}\right| = \frac{\beta_1 C}{\beta_3} \tag{14}$$

According to Formula (14), when $C=0$, the steady-state error approaches 0; meanwhile, the derivative of angular acceleration d is 0. That is, when the speed changes with the acceleration unchanged, ESO-PLL can carry out errorless tracking of the speed.

4. Analysis of experimental simulation results

The effectiveness of the proposed method is verified by simulation examples. The simulation parameters are given as follows. The power rating is given as 4.5 kW; the rated speed is given as 1000 r/min; the voltage is given as 380 V; the current is given as 7.6 A; the stator winding resistance is given as 1.7; the *d*- angle inductance is given as 0.0045 H; the *q*-axis inductance is given as 0.015 H; the flux linkage is given as 0.1827 Wb; the pole pairs is given as 4. Furthermore, the vector control method is adopted in simulation to collect the data of PMSM running time at $i_d=0$ and $i_d=-2$, respectively. The sampling time is 1 s, and the sampling frequency is 0.0001. The population size is 50 and is divided into 5 groups with 150 iterations. Figures 4-7 show the identification result curves.

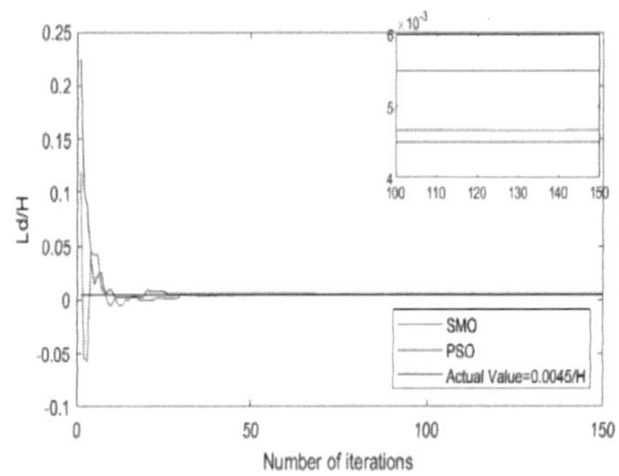

Figure 4. L_d identification curve

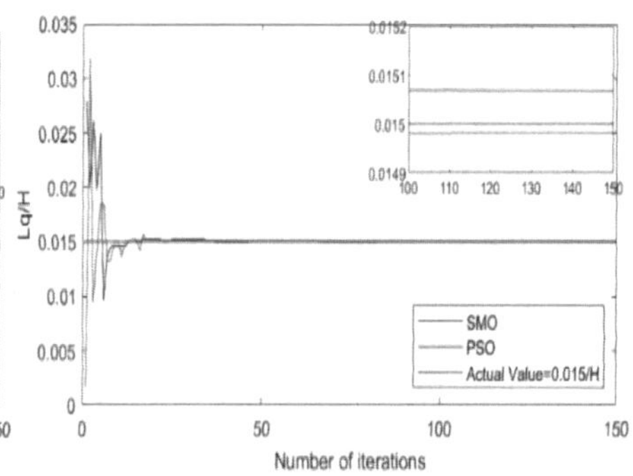

Figure 5. L_q identification curve

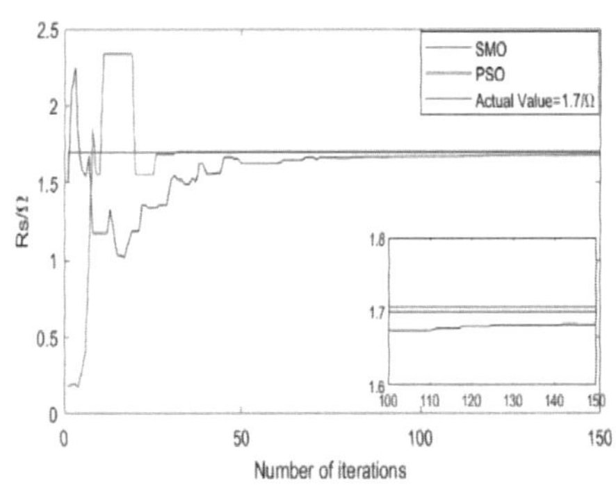

Figure 6. R_s identification curve

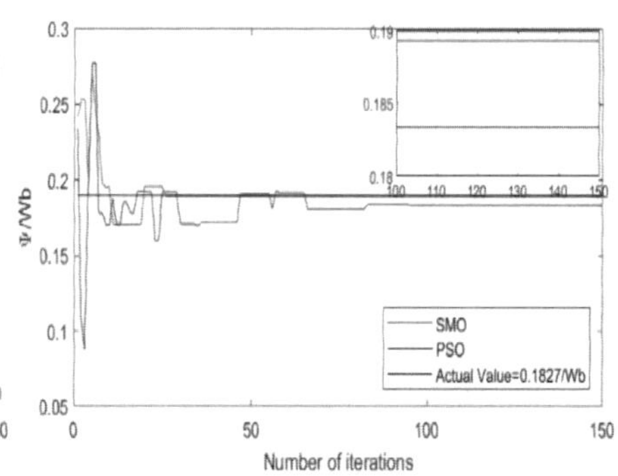

Figure 7. ψ_f identification curve

Simulation results show that the SMO is significantly better than the PSO, and the optimisation speed is faster. Figure 8 gives the fitness function curve. Figure 8 shows that both algorithms can converge to the true identification value of PMSM parameters and is not falling into the optimal local solution. However, the SMO algorithm converges significantly faster with higher accuracy.

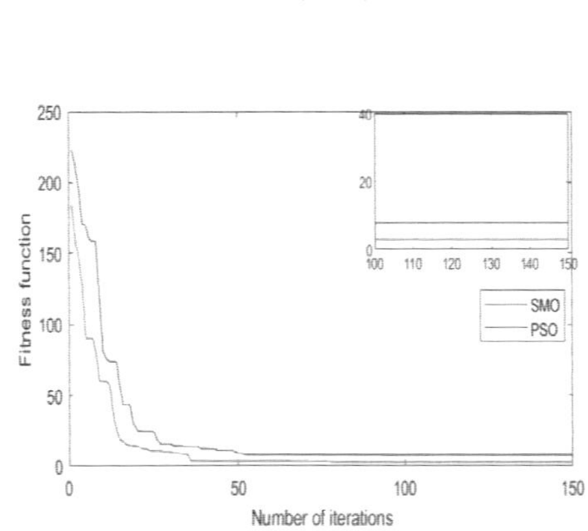

Figure 8. Fitness function curve

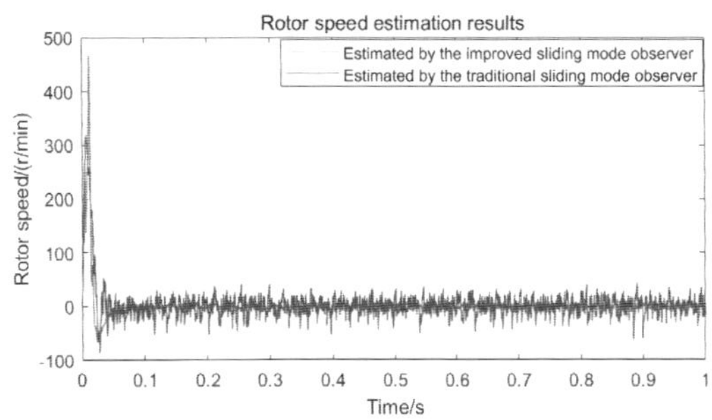

Figure 9. Rotor speed estimation results by the two observers

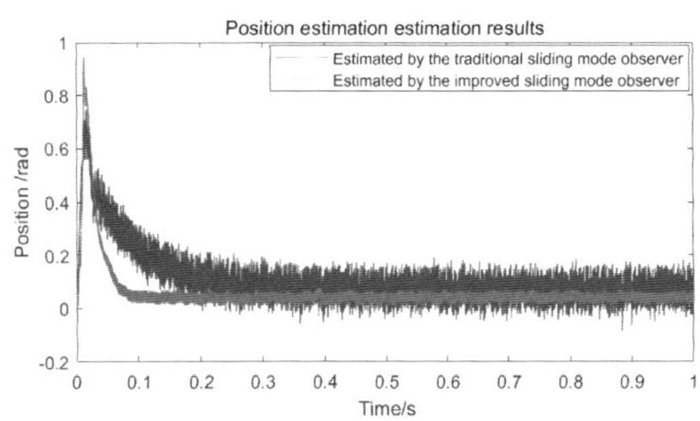

Figure 10. Position estimation by the results of the two observers

Figures 9 and 10 show the comparison results of the rotor position and velocity estimated by the SOBS before and after the modification. From Figures 9 and 10, it is seen that the rotor position estimation error of the traditional SOBS is larger when the motor starts. On the contrary, the improved SOBS reduces the obvious chattering problem in the low-speed range. The simulation results fully illustrate that the improved SOBS is more accurate and robust than the traditional SOBS in speed tracking.

5. Conclusions

This paper studied the parameter identification of PMSM and the vector control algorithm without a position sensor. By analysing the change in the sensorless control performance of the motor, the SMO switch function was modified, and the parameter identification results based on the spider monkey algorithm were introduced into sensorless control. The rotor position and velocity were estimated by the ESO-PLL. Simulation results showed that the proposed method effectively improved the sensorless control performance of PMSM and tracked rotor position and speed more accurately.

The possible improvements include adding adaptive methods to improve the SMO algorithm and reducing harmonic effects and position compensation to improve the precision of the SOBS.

References

[1] Chen G, Xu L. Adaptive quasi-proportional resonant control with parameter estimation for PMSM sensorless control[J]. Advances in Mechanical Engineering, 2023, 15(2): 16878132231154069.
[2] Li Y, Hu H, Shi P. A Review of Position Sensorless Compound Control for PMSM Drives[J]. World Electric Vehicle Journal, 2023, 14(2): 34.
[3] Yu Y, Huang X, Li Z, et al. Full parameter estimation for permanent magnet synchronous motors[J]. IEEE Transactions on Industrial Electronics, 2021, 69(5): 4376-4386.
[4] Chen R B, Cao T Q, Guo X Y, et al. Speed Sensorless Control of Permanent MagnetSynchronus Motors Based on MRAS[J]. Electrical Measurement & Instrumentation, 2021, 58(8): 179-184.
[5] Chen Z, Zhang X, Zhang H, et al. Adaptive sliding mode observer-based sensorless control for SPMSM employing a dual-PLL[J]. IEEE Transactions on Transportation Electrification, 2021, 8(1): 1267-1277.
[6] Zhao K, Yang L, Shuang Z, et al. Improved Sensorless Control of Permanent Magnet Synchronous Motor Based on Two-stage Filter[C]//2022 25th International Conference on Electrical Machines and Systems (ICEMS). IEEE, 2022: 1-6.
[7] Guo X, Yin Z, Zhang Y, et al. Position sensorless control of PMLSM based on adaptive complex coefficient sliding mode observer[J]. Energy Reports, 2022, 8: 687-695.
[8] Li T, Zhao Y, Hou L. Adaptive Sliding Mode Control With Disturbance Observer for Speed Regulation System of Permanent Magnet Synchronous Motor[J]. IEEE Access, 2023, 11: 17021-17030.
[9] Chen T, Chen B. Nonlinear Identification of PMSM Rotor Magnetic Linkages Based on an Improved Extended Kalman Filter[J]. Journal of Sensors, 2022, 2022.
[10] Sarkar A. Neural coordination through spider monkey optimization-guided weight synchronization[J]. Multimedia Tools and Applications, 2023: 1-30.

Flight state recognition for UAV optical flow velocity measurement

Yu Gong *[a], Xiaohong Liu [a]

[a] Yancheng power supply company State Grid Jiangsu Yancheng, China;

* Corresponding author: 1780462535@qq.com

Abstract. For the visual navigation and positioning of drones, the optical flow algorithm is used to calculate the flight speed of the drone, the calculation errors of its rotation state, and the vertical take-off and landing state. Based on the characteristics of its optical flow data, this paper compares various classification calculation methods. It uses the median KNN curve of the optical flow vector to simplify its feature quantity. Finally, it realizes the motion of the UAV with the optical flow characteristics based on the SVM algorithm. State recognition experiments have proved that this method can effectively identify the motion state of the UAV, and the calculation efficiency is high.

1. Introduction

In recent years, the application of drones has increased. Compared with traditional manned aircraft, multi-rotor drones have the advantages of low cost, small size, and easy use. In places unsuitable for manned aircraft to approach and shoot, drones can replace manned aircraft to complete the shooting task. At present, most UAVs have a vision system, and the optical flow method is an important research direction of UAV vision in terms of target tracking, positioning, and speed measurement [1].

There are two methods for speed measurement using images, feature matching and feature tracking, both of which must first detect feature points and then process them. However, since feature matching uses a one-to-many method to match feature points and calculates the degree of similarity to obtain the best matching point, compared with the one-to-one method of feature tracking, the amount of calculation will undoubtedly increase [2].

In 2013, ETH Zurich studied an optical flow sensor called PX4Flow, which is widely used in UAVs to calculate the speed information of UAVs [3]. In 2016, Andrii et al. [4] at Kharkiv National Aerospace University in Ukraine developed a small UAV obstacle detection system based on the optical flow method. In 2019, Tien et al. [5] from the University of Minnesota proposed a parameter adaptive optical flow algorithm based on the characteristics of optical flow data, which improved the performance of the optical flow algorithm in calculating the horizontal velocity of the drone. In 2021, Negoro et al. [6] proposed a flight speed measurement method based on image processing to calculate the energy model of the UAV. In 2021, Sibilski et al. [7] developed an optics-based system to estimate the ground speed of a drone in the event of GNSS failure, jamming, or unavailability.

In 2014, Lu Qiang et al. [8] installed an optical flow sensor at the bottom of the quadrotor aircraft. They used the optical flow information to detect the horizontal motion speed of the quadrotor aircraft relative to the ground. In 2022, Wei [9] of the North University of China proposed a speed measurement method based on a square root volumetric Kalman filter to realize the image optical flow speed measurement of UAV.

The above-mentioned domestic and foreign research methods use the optical flow method to track the feature points to measure the visual velocity of the UAV. Still, none of them consider the optical flow calculation problem in the non-translational state. Therefore, this paper proposes a UAV visual motion state classification method based on the SVM algorithm to solve the problem of wrong calculation of the optical flow velocity in the rotation and lifting state during the UAV visual navigation and positioning process.

2. Optical flow velocity calculation

The method proposed in this paper is to use the onboard camera of the UAV to be connected to the UAV. During the movement of the UAV, the image background changes, and the optical flow method is used to calculate the displacement of the pixel movement in the image. According to the camera, the focal length and the height from the ground calculate the projection relationship of the pixels, and the speed information is calculated according to the frame rate.

2.1. Surf feature point detection

In this paper, the grayscale processing of the image is performed first, and the feature point detection is performed on it. The Surf feature descriptor is selected for optical flow tracking. The Surf feature point is developed from the Sift [10] feature point. Compared with the Orb [11] feature point, the Sift feature point is Fast [12], and the image pyramid model is introduced. After the image is sampled up and down, a Gaussian pyramid is created. The difference calculation is performed on the same scale. Then feature point detection is performed so feature points can be accurately detected at multiple scales. The feature descriptor is added to calculate the main direction of the feature in the area, which has a strong Anti-rotation. The disadvantage is that it is time-consuming to detect feature points. After preserving the advantages of Sift[13], its calculation speed is accelerated and is about an order of magnitude faster than Sift in terms of time-consuming feature point detection.

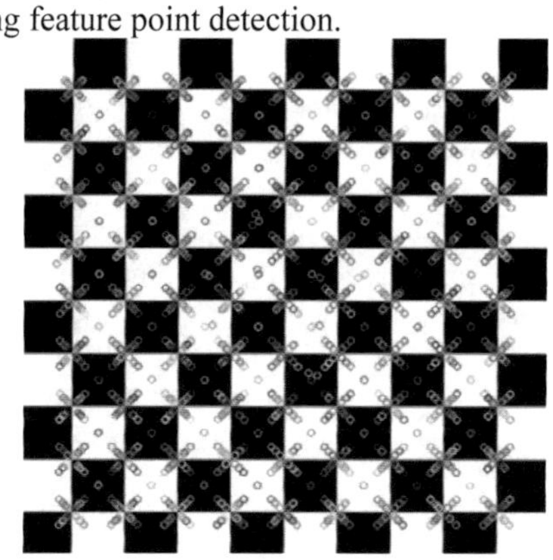

Figure 1. Surf feature point detection map

Table 1. The average time-consuming of feature point detection.

Feature description	Average time spent/ms
Sift	55/260=0.212
Surf	43/1549=0.028
Orb	25/500=0.05
Fast	17/302=0.056
Shi-Tomas	20/90=0.222

Table 1 shows the detection time of a single feature point of various algorithms for feature point detection on the same image. In this paper, the Surf feature descriptor is used for optical flow tracking, and the Surf feature points are shown in Figure 1.

2.2. Optical flow field and velocity field conversion

After the feature points are tracked, the pixel displacement information of the feature points is obtained, and the real speed information can only be obtained by changing the coordinates.

Honegger et al.'s study [14] gives the relationship between the speed of the UAV and the speed of optical flow under the condition of shooting on the ground:

$$\begin{cases} T_x = \dfrac{v_x Z}{f} + \omega_y Z \\ T_y = \dfrac{v_y Z}{f} - \omega_x Z \end{cases} \tag{1}$$

In Equation (1), T is the real motion speed of the drone, v is the calculated optical flow velocity value, Z is the height of the drone to the ground, f is the focal length of the camera, and w is the angular velocity of the drone.

The angular velocity component in Equation (1) has nothing to do with the height of the UAV but only the attitude of the UAV [15]. Therefore, the movement of the UAV between two frames can be approximately converted into a translational movement. The rotation part can be considered as zero, and then it can be simplified to obtain the linear motion velocity expression::

$$\begin{cases} T_x = \dfrac{v_x Z}{f} \\ T_y = \dfrac{v_y Z}{f} \end{cases} \tag{2}$$

3. Optical flow feature analysis

In calculating the optical flow vector, we found that when the UAV rises and falls in a straight line, the optical flow vector data will still be calculated. The data is extremely unstable, which cannot represent the real speed at that time at all. For example, when the UAV rotates, its height remains unchanged, and its speed is zero, but the optical flow data will still be calculated, which will cause a large error. In addition, when the UAV ascends or descends, the ascending and descending speed can be calculated only based on the height information. The introduction of optical flow information causes a huge error to it. It is only based on the attitude information of the acquired image and the speed calculation equation. It is also impossible to know whether the current motion state is a flight state, a lifting state, or a rotating state. Since the optical flow value is processed, it should be analyzed and identified from the data perspective. Its optical flow calculation value is set to zero.

We first analyze the optical flow value:

1. Rotation state: When the angular velocity of the image is constant, the farther it is from the center, the larger its motion scale will be. Correspondingly, the motion scale of the detected feature points will also be larger, and the optical flow value calculated after correct tracking will also be larger. Moreover, the correct optical flow vector should be perpendicular to the line connecting the center point and the feature point at the previous moment. The optical flow value should rotate in the same direction.

2. Rise and fall state: Rise or fall will make the image zoom in or out, which is also carried out around the center point. The correct data direction it detects is shrinking or expanding toward the center point, and its size is also farther away from the center point. The farther it is away, the greater the correct optical flow value will be.

3. Flight state: From the analysis in the previous chapter, we can see that the correct direction of the optical flow vector measured in the flight state should be consistent with the magnitude of the optical flow value.

Based on the above analysis of the characteristics of optical flow data, we can use its data characteristics to identify its operating mode and then calculate the optical flow velocity value according to the motion state.

As shown in Figures 2-4, the distribution of optical flow vectors on a picture of 500*500 pixels is calculated for the rotation state, lift state and flight state.

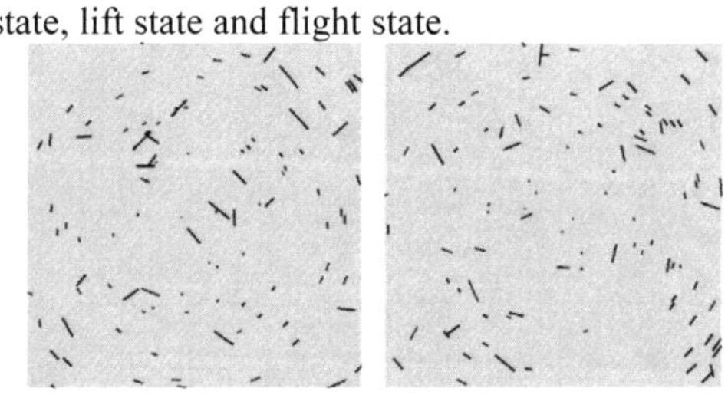

Figure 2. Lift state

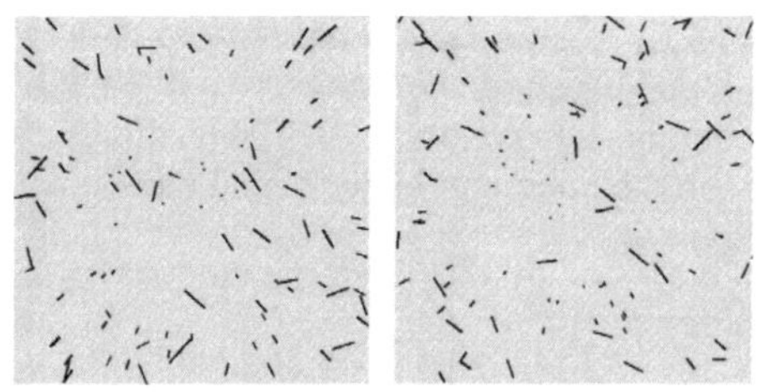

Figure 3. Rotation state

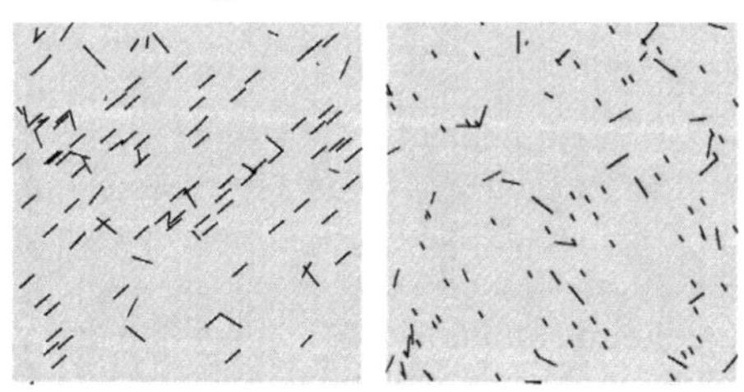

Figure 4. Translational state

In this paper, 1000 pictures are collected for the three states. The training set is 60%, the test set is 20%, and the verification set is 20%. 20 epochs are set, and the loss rate is set. The epoch is with the highest accuracy. This paper uses the classic Resnet artificial neural network and Resnet18 transfer learning [16] to find that in the self-built network and the image classification of the above three states. When the parameter amount reaches 4.77 Mb, the classification accuracy of the test set is 87.5% and 88.3% in the test set, which still has a certain error. The classification speed is an average of 57.12 ms per image (Figures 5-6 show the training loss function data diagram and the training accuracy data diagram). After using Resnet18 migration learning, the classification accuracy rate of the test set is 97.8%, reaching 97.5% in the test set. Still, its parameter volume has also reached 42.7 Mb. At this time, the classification speed of motion vector graphics is very low. It has also reached 0.2 s per frame. For the motion scene of the drone, the calculation amount is more complicated, and the classification speed is also low.

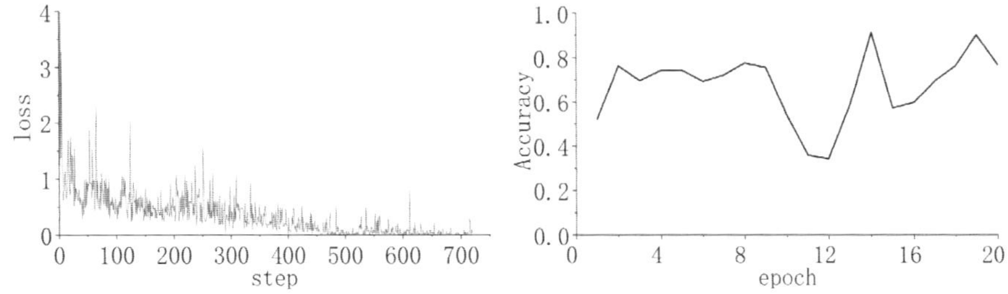

a) Resnet training loss function curve b) Resnet training accuracy curve

Figure 5. Artificial Resnet network training curve

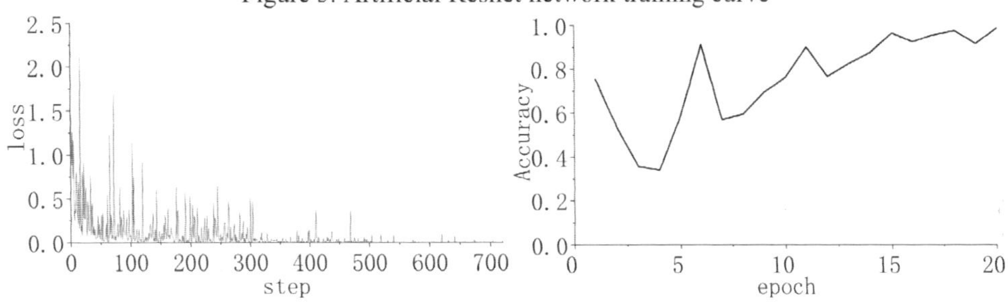

a) Resnet training loss function curve b) Resnet training accuracy curve

Figure 6. Resnet18 network training curve

4. Optical flow feature simplification

It can be seen from the above motion vector diagrams that the distribution of optical flow vectors in the three motion states is obvious. Still, two-dimensional image classification needs to calculate a series of calculations, such as convolution and pooling, for it can be a simplified data type.

It can be observed from the above Figures 2-4 that the size distribution of the optical flow vector is different. The value distribution of the lift state and the rotation state is uniform, and the value of the translation state is relatively concentrated. This article classifies the KNN curves that use the size changes of these data in advance. Although there will inevitably be some tracking error points during motion, most of the values of the optical flow vector represent the values of the correct motion state. Although the size and number of data are unknown, analysis shows that such data account for a small proportion of the global data. With careful consideration, this paper chooses the median as the initial representative value of the movement speed of the UAV. Based on this value, the offset of all optical flow values relative to the x-axis and y-axis of this value is calculated. The total offset is then calculated and sorted from small to small. The distance between each feature point and the median of the data is obtained by calculating the offset in the xy direction. After analysis, it can be seen that the data distribution of the lifting state and the rotating state are basically the same, and the state characteristics are shown in Figures 7-9:

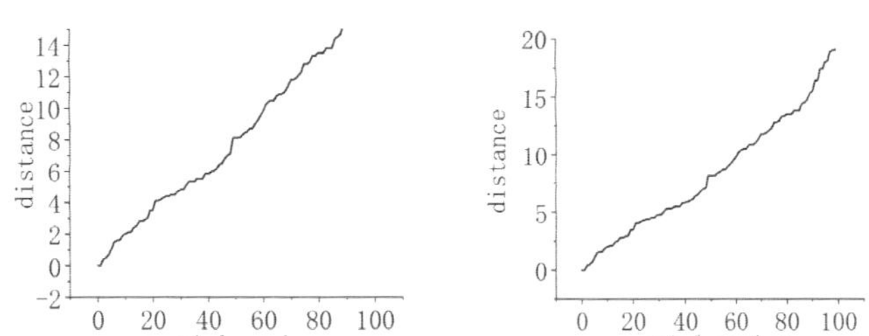

Figure 7. Lift state

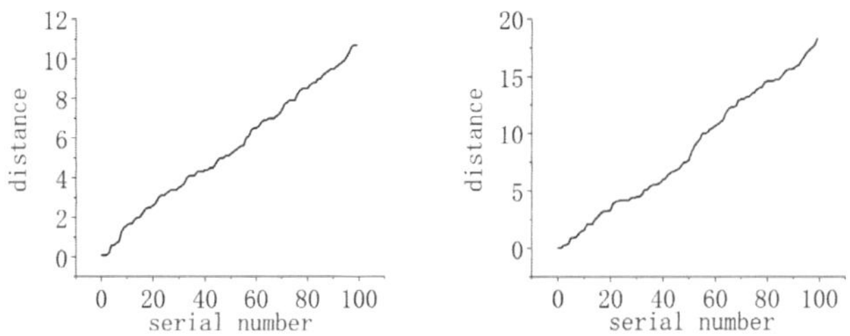

Figure 8. Rotation state

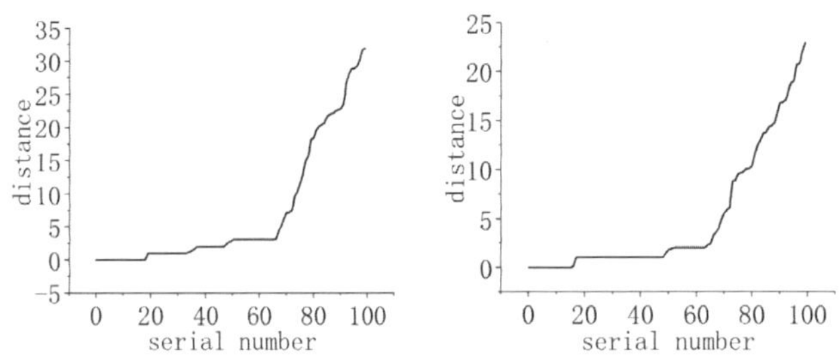

Figure 9. Translational state

It can be observed from Figures 7-9 that since most of the background optical flow represents the motion state of the UAV, the median of the optical flow vector offset in the translational state basically represents the motion offset of the UAV. There is a large part of the value and other offsets, which are all at a relatively small value, and then there will be a large jump. As for the rotation and lifting state, since the offset representing the movement state of the UAV is related to the position of the feature point in the image, the median value of its offset is basically in the middle of all offsets. Therefore, the difference with other offsets is more evenly distributed, and its curve is closer to a straight line. Through such a data processing method, the dimensionality of the data is also reduced, and the computational complexity of the algorithm is also greatly reduced.

5. SVM classification

5.1. SVM algorithm

Support Vector Machines (SVM) is a commonly used classification method in machine learning. It is often used to deal with small sample problems of binary classification, and it has better processing results for nonlinear problems. The main idea is to find the hyperplane that maximizes the interval so that the difference between the samples on both sides of the plane is as large as possible, thereby improving the generalization ability of the model [17].

There are two main types of data classification, namely linearly separable data and linearly inseparable data. For linearly separable, it can be classified by constructing a linear classifier. On the contrary, a nonlinear classifier is constructed by introducing a kernel function to map linear inseparable data into high-dimensional space, using the quadratic programming problem to solve the optimal decision hyperplane[18]. Figure 10 shows the two-category discriminant model of the SVM algorithm..

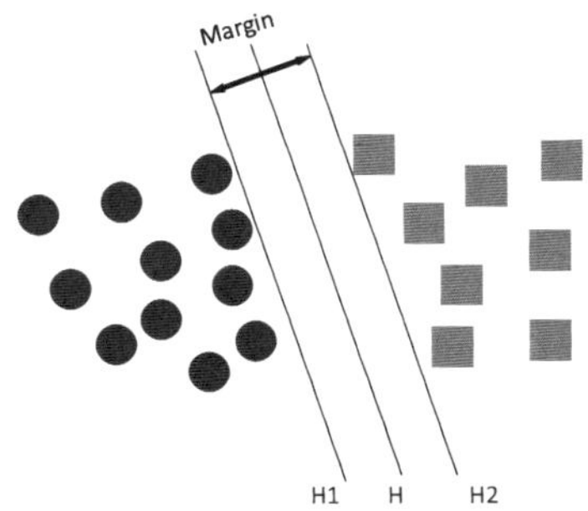

Figure 10. SVM binary classification discriminant model

The hyperplane divides the two types of samples. When the distance between the sample and the two sides of the plane is the largest, the model has the strongest noise tolerance for the sample. The samples in the dotted line become "support vectors" and together determine the location of the decision plane.

The optimization problem of SVM has the following form:

$$\begin{cases} \min\limits_{w,b} \frac{1}{2} \|w\|^2 = \min\limits_{w,b} \frac{1}{2} w^T w \\ s.t. y_i(w^T x_i + b) \geq 1, i = 1,2,\cdots,n \end{cases} \tag{3}$$

In Equation (3), (w, b) are the slope and intercept of the straight line, and (x_i, y_i) are the coordinates of the i-th point.

In general, the data is not a linearly separable data set. At this time, it is necessary to introduce a kernel function to convert it into a linearly separable data set. When a few samples are allowed to make mistakes, the slack variable g_i is introduced. At this time, the objective function can be expressed for:

$$\begin{cases} \min\limits_{w} (\frac{1}{2} \|w\| + c \sum_{i=0}^{n} g_i)^2 \\ s.t. y_i(w^T \varphi(x_i) + b) \geq 1 - g_i, g_i \geq 0, i = 1,2,\cdots,n \end{cases}. \tag{4}$$

The c in the above Equation (4) is the penalty factor, and the numerical value of the penalty factor c represents the penalty for misclassification. The larger the value is, the more attention is paid to the wrong classification. However, it will also cause the generalization ability to decline. Since the slack variable is added, a new Lagrangian multiplier μ_i should be introduced. The corresponding Lagrangian function is:

$$L(w, b, \alpha, g, \mu) = \frac{1}{2} \|w\|^2 + c \sum_{i=0}^{n} g_i + \sum_{i=0}^{n} \alpha_i (1 - y_i(w^T \varphi(x_i) + b) - \sum_{i=0}^{n} \mu_i g_i). \tag{5}$$

The solutions of SVM are all globally optimal, but the selection of kernel function and penalty factor determines the performance of SVM classification.

5.2. SVM algorithm parameter determination

Common kernel functions include the Gaussian kernel function, polynomial kernel function, and linear kernel function. Due to the variety of UAV scene changes, its data is also changeable, and linear kernel SVM is generally used for multi-classification tasks (greater than or equal to 3) tasks. Non-linear kernels are mostly used in binary classification tasks. Hence, this paper uses the most widely used Gaussian kernel function, which has better performance when there are not many samples, and its function expression is shown in Equation (6):

$$K(x_i, x_j) = \exp\left(-\frac{\|x_i - x_j\|^2}{\delta^2}\right). \tag{6}$$

In this paper, 1000 groups of the two types of states are collected for fitting, and all the distance features are simplified to 100 data points. For those less than 100, the index is randomly selected to insert it. The inserted value is the average value of the left and right sides. For more than 100 For the data of 100 data points, the data is randomly selected within the data for deletion, all of which are simplified to the characteristics of 100 data points for analysis. Finally, 400 sets of data are used for accuracy testing.

In this paper, when the penalty factor C=0.01 is set, the running accuracy and calculation time of the three kernel functions are shown in Table 2 below.

Table 2. The results of different kernel functions

	Accuracy	Sample run time
Polynomial kernel function	82.8%	100 ms
Gaussian kernel function	93.35%	253 ms
Sigmod function	76%	463 ms

The classification accuracy and running time of different penalty factors of the Gaussian kernel function are shown in Table 3:

Table 3. The results of different kernel functions

Penalty Factor	Accuracy	Running Time
0.01	93.35%	253 ms
0.1	98.5%	60 ms
1	100%	25 ms

From the above analysis, we can see that this paper uses an SVM support vector machine to classify its flight status, the kernel function uses Gaussian kernel function, and the penalty factor uses c=1.

6. Conclusion

This paper starts from the principle of visual speed measurement and expounds on the advantages and disadvantages of the feature point detection algorithm. It analyzes the conversion relationship between the motion field and the velocity field and simplifies it. Then, it aims at the calculation error of the optical flow vector of the UAV rotation and lifting state. It analyzes the problem, extracts its optical flow vector diagram, and uses the classic image classification algorithm residual network to classify. It is found that the image-based classification algorithm has a slight disadvantage in computational efficiency. Therefore, its feature state is simplified, and the optical flow is analyzed. The KNN distance curve of the median is calculated. Then, the SVM algorithm is used to classify it. Its kernel function and penalty factor are determined. The classification efficiency is high, and the calculation speed is also very fast, which is suitable for visual speed measurement and navigation Position any scene.

Acknowledgment

The authors thank the anonymous referees for their helpful comments and suggestions. The project was supported by Science and Technology Project of State Grid Jiangsu Electric Power Co., Ltd. (State Grid Yancheng Power Supply Company) (No.J2021041).

References

[1] Yang Shengwei, Zhao Wei, Liu Jianye. Optical flow speed measurement optimization method for UAV navigation [J]. Electro-optic and Control, 2019, 26(1):5.
[2] Yang Shengwei. Research on Optimization Algorithm of Optical Flow Velocity Measurement of UAV [D]. Nanjing University of Aeronautics and Astronautics, 2019. DOI: 10.27239/d.cnki.gnhhu.2019.000756.

[3] Honegger D, Meier L, Tanskanen P, et al. An open source and open hardware embedded metric optical flow CMOS camera for indoor and outdoor applications[C]// IEEE International Conference on Robotics & Automation.IEEE,2013.
[4] Molchanov A, Kortunov V, Mozolyuk S. A real-time system for detection and avoidance of dynamic obstacles for micro-UAVs[C]// International Conference on Methods & Systems of Navigation & Motion Control. IEEE, 2016.
[5] Tien D, Carrillo-Arce L C, Roumeliotis S I. High-speed autonomous quadrotor navigation through visual and inertial paths[J]. The International Journal of Robotics Research, 2018:027836491878657.
[6] Ito S, Negoro S, Kong X, et al. A Methodology for Measuring Flight Speed of Drones in Indoor Environments [J]. Procedia Computer Science, 2021, 187(1):322-328.
[7] Sibilski K. Ground Speed Optical Estimator for Miniature UAV[J]. Sensors,2021,21.
[8] Lu Qiang, Ni Peipei, Wang Guosheng, Liu Feng. Hover correction of quadrotor aircraft based on optical flow sensor [J]. Journal of Armored Forces Engineering College, 2014,28(03):68-72.
[9] Wei Jiaqi. Autonomous speed measurement and positioning technology based on improved optical flow method [D]. North University of China, 2022.DOI: 10.27470/d.cnki.ghbgc.2022.000500.
[10] Lowe D G. Distinctive Image Feature from Scale-Invariant Key points[J]. International Journal of Computer Vision, 2004.
[11] Rublee E, Rabaud V, Konolige K, et al. ORB: an efficient alternative to SIFT or SURF[C]// IEEE International Conference on Computer Vision, ICCV 2011, Barcelona, Spain, November 6-13, 2011. IEEE, 2011.
[12] Rosten E. Machine learning for very high-speed corner detection[J]. ECCV'06, May, 2006.
[13] Bay H, Tuytelaars T, Gool L V. SURF: Speeded up robust features[C]// Proceedings of the 9th European conference on Computer Vision - Volume Part I. Springer-Verlag, 2006.
[14] Honegger D, Meier L, Tanskanen P, et al. An open source and open hardware embedded metric optical flow CMOS camera for indoor and outdoor applications[C]// IEEE International Conference on Robotics & Automation. IEEE, 2013.
[15] Yuan Ruyue. Research on FPGA-based Optical Flow Positioning System for UAVs [D]. Harbin Institute of Technology, 2017.
[16] Zhao Chunxia. Application of Image Classification Based on ResNet18 in Diagnosis of Crop Diseases and Pests [J]. Agriculture and Technology, 2021,41(19):10-13. DOI: 10.19754/j.nyyjs.20211015003.
[17] Xia Jingwen. Application Research of Support Vector Machine (SVM) in the Classification of Ancient Ceramics [D]. Jingdezhen Ceramic University, 2022. DOI: 10.27191/d.cnki.gjdtc. 2022.000296.
[18] Xu Jie. Research on the Effectiveness of Industrial Investment Based on SVM Model [D]. Chongqing Technology and Business University, 2022.DOI: 10.27713/d.cnki.gcqgs. 2022.000108.

AUTHOR INDEX

Institute of Physics
Dirac House, Temple Back
Bristol BS1 6BE UK

ISSN: 1742-6588
ISBN 978-1-7138-7759-2

9 781713 877592

CURRAN ASSOCIATES INC.
proceedings.com